Introduction to Statistics Using Excel

Kazumitsu NAWATA

Asakura Publishing

First published in Japan in 2021
by Asakura Publishing Company, Ltd
6-29, Shin'ogawa-machi, Shinjuku-ku,
Tokyo 162-8707, Japan

Copyright © 2021 by Kazumitsu Nawata

ISBN 978-4-254-12262-6

All rights researved. No part of this publication may be reproduced, stored in a retrieval system,
or transmitted in any form or by any means, electronic, mechanical, photocopying, recording or otherwise,
without prior written permission.

Preface

In statistics, many data analysis methods and theorems have been developed, and many textbooks explaining them have been published. It is very important to analyze actual data sets using computers. This would also help people to understand these methods and theorems. Fortunately, today, high-powered personal computers (PCs) and PC programs are available at relatively low cost, and it is possible to analyze most data sets using them.

In this textbook, we will learn how to do data analyses using Microsoft Excel based on the lectures I have been giving at the University of Tokyo. Excel is one of the most popular programs in the world. It is easy to use and is widely used in fields other than statistics, and it is a beneficial tool in various fields. Since this textbook shows how to do exercises step by step, it can help promote the understanding of statistical methods and theorems. When we use statistical software packages such as Eviews, R, SAS, and SPSS, we can obtain results without any knowledge of statistics. I believe this textbook will be beneficial as an introduction to statistics and data analysis.

I have written this textbook according to the following principles for those who are beginners in statistics and PCs.

i) I only explain the basics of statistical methods and theorems, and I emphasize how to perform data analysis using Excel. If readers are interested in the theoretical and methodological details, please read other statistical textbooks such as those listed in (**Statistical Theories and Methodologies**) of References.

ii) I only explain the basic features of PCs, Windows, and Excel and avoid unnecessary complications. For example, if the same results can be obtained by multiple different operations, I only explain one method for the sake of simplicity and usefulness. This textbook covers both Excel and statistics so that readers can get enough basic knowledge and skills to perform data analysis. If readers are interested in more exercises, please read textbooks listed in (**Excel, Data Analysis**) of References.

We will mainly learn how to treat and summarize the data from Chapter 1 to

Chapter 8. These chapters include few theoretical parts, so this book can be used as an introductory textbook on information analysis and Excel. This book uses the Microsoft 365 version of Excel (as of August, 2020) on Windows 10, but the results are also checked in Excel 2016.

This book summarizes the contents of my lectures given at the University of Tokyo. The first edition was written in 1996 in Japanese, and three major revisions have been done over 20 years. This book is an English edition of the 4th Edition, the most recent one, published in 2020. (However, it is not just a simple translation.) I would like to thank the Editorial Office of the Asakura Publishing for their kind consideration in publishing this book in English.

July 2021

Kazumitsu NAWATA

Graduate School of Engineering,
University of Tokyo,
and
Research Center for Health Policy and Economics,
Hitotsubashi Institute for Advanced Study,
Hitotsubashi University

The data of Table 4.1 are available from the website of the Asakura Publishing. Access (https://www.asakura.co.jp/books/isbn/978-4-254-12262-6/) and go to this book's page.

Windows10, Microsoft 365, Excel, and Word are trademarks or registered trademarks of Microsoft Corporation in the United States of America and other countries throughout the world. The other company and product names that appear in this book are trademarks or registered trademarks of the companies. I do not explicitly include the TM and other marks in the text.

Contents

1

Introduction to Excel

1.1 How to Use Excel

Here, we learn how to use Excel. To distinguish the explanatory passages from what is actually input into Excel, the words and commands that we actually input into Excel are printed in **bold**. (This is only for clarity; we don't use bold characters when we use Excel.) For readers who are interested in more details, see the textbook in (**Excel, General**) of References.

1.1.1 How to Start Excel

To start Excel in Windows 10, click [Start] → [Excel] (Fig. 1.1). Using the mouse, move the arrow (this is called the "mouse pointer") to [Start] and click the left mouse button. We only use the left button in this book. After Excel starts, click [New] → [Blank workbook]. An Excel "workbook" appears on the screen (Fig. 1.2). The names of its components are given in Fig. 1.3.

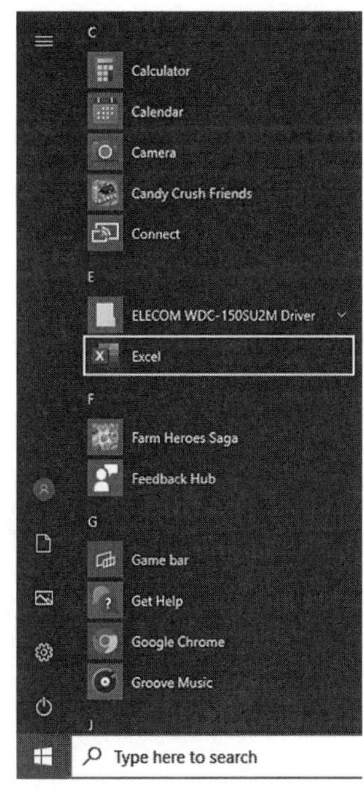

Fig. 1.1 Start Excel by clicking [Start] → [Excel].

1.1.2 About Ribbon

In Excel (and other Microsoft Office products such as Word), the interface called the "Ribbon" is used. The commands are divided into [Tab] and [Group]. [Tab] is displayed on the upper side of the Ribbon, including [File], [Home], [Insert], [Page Layout], [Formulas], [View] and so on. The Ribbon changes depending on [Tab]. When we initially start Excel, the [Home] tab is selected, and commands listed under other tabs are not displayed. If we want to use commands in another tab, check that tab.

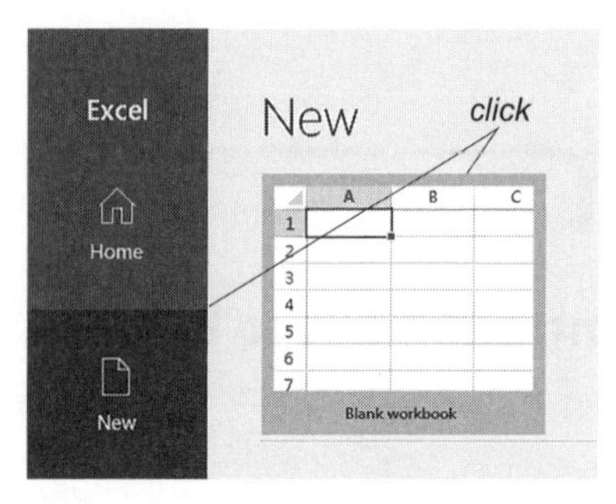

Fig. 1.2 Click [New] → [Blank workbook].

In the selected tab, commands are divided into [Group] based on the similarities among operations. The [Home] tab includes [Clipboard], [Font], [Alignment], [Number], [Styles], [Cells] and [Editing] (Fig. 1.4).

In [Clipboard], you will find the command buttons [Paste], [Cut], [Copy] and [Format Painter] (Fig. 1.5). This is similar to finding a residential address by searching the names of countries, states or provinces and cities. It is necessary to understand the structure of the Ribbon for make the best use of Excel.

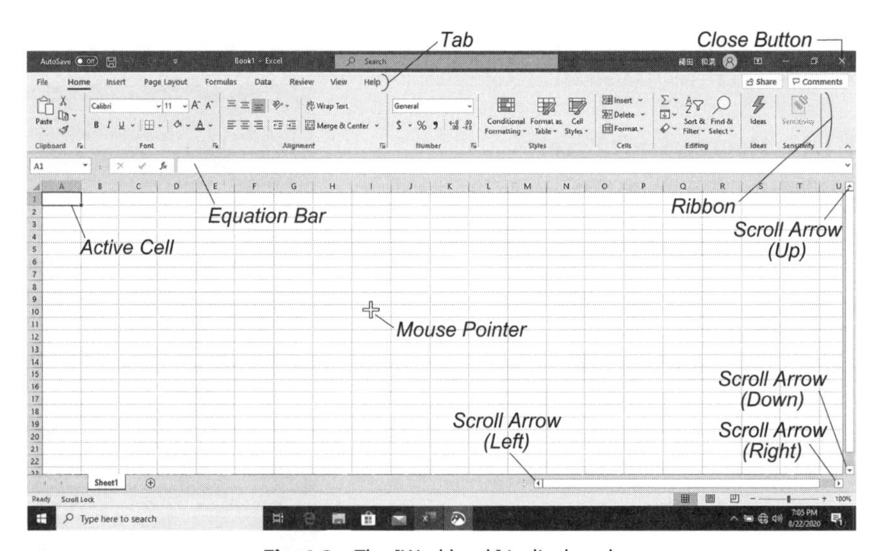

Fig. 1.3 The [Workbook] is displayed.

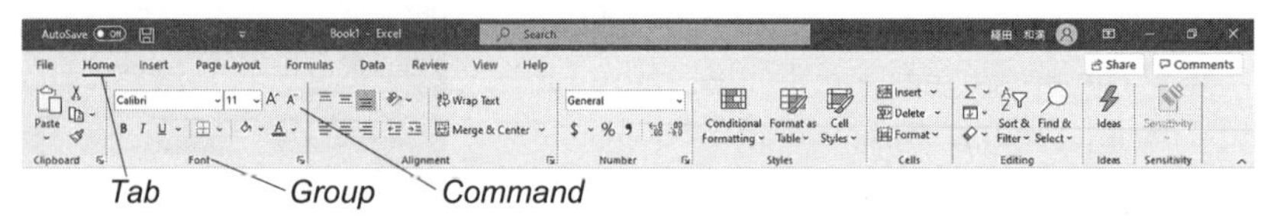

Fig. 1.4 Several tabs are shown. The Ribbon changes depending on which tab is clicked. The [Home] tab includes seven groups.

1.1.3 Inserting Data Analysis

To do statistical analysis using Excel, we need to add the [Data Analysis] tools to Excel. Since we do not use [Data Analysis] until Chapter 5, you may do it later if you are not familiar with Excel. After starting Excel, select the [File] tab by clicking on it using the mouse. The command groups in the [File] tab are displayed. Then, select [Option] by clicking on it (Fig. 1.6). The [Excel Options] box appears. Select

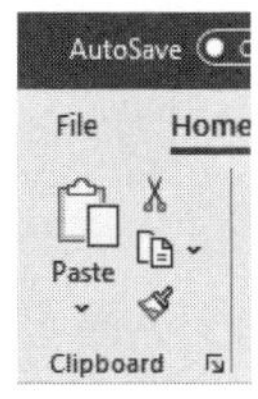

Fig. 1.5 In [Clipboard], you will find the command buttons [Paste], [Cut], [Copy], and [Format Painter].

Fig. 1.6 To insert [Data Analysis], select the [File] tab → [Option]. If [Options] does not appear, click [Move...].

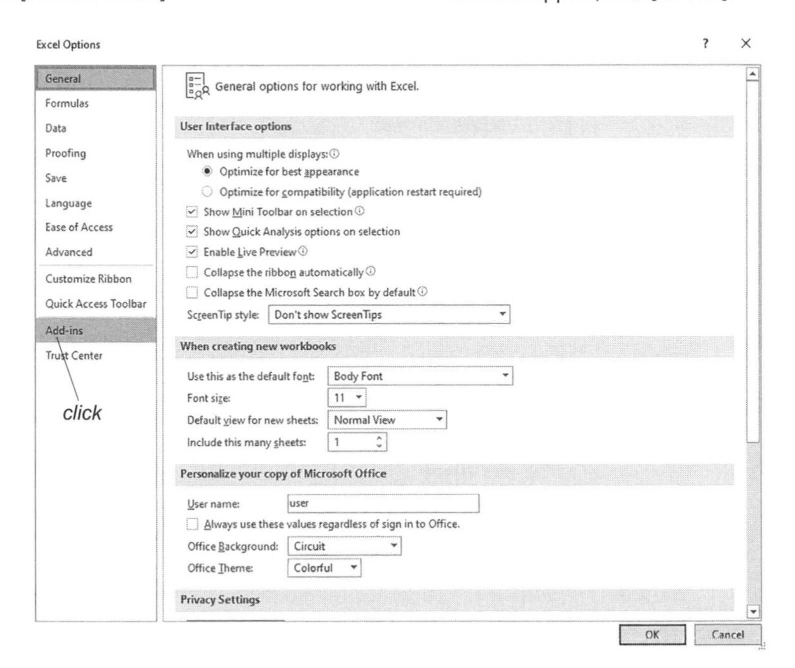

Fig. 1.7 The [Excel Options] box appears. Select [Add-in] on the left side of the box.

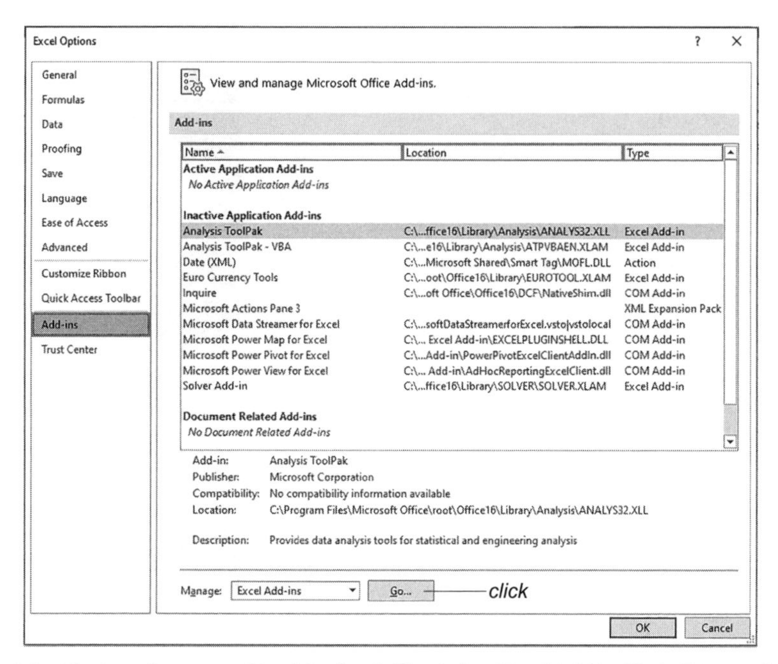

Fig. 1.8 The box changes to [Excel Options]. Check that [Excel Add-ins] is in [Manage] and click [Go].

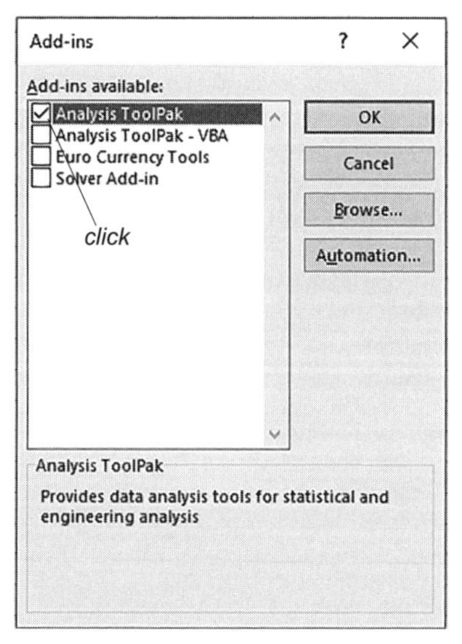

Fig. 1.9 The [Add-ins] box appears. Then click the checkbox in front of [Analysis ToolPak] so that it is checked, and click [OK].

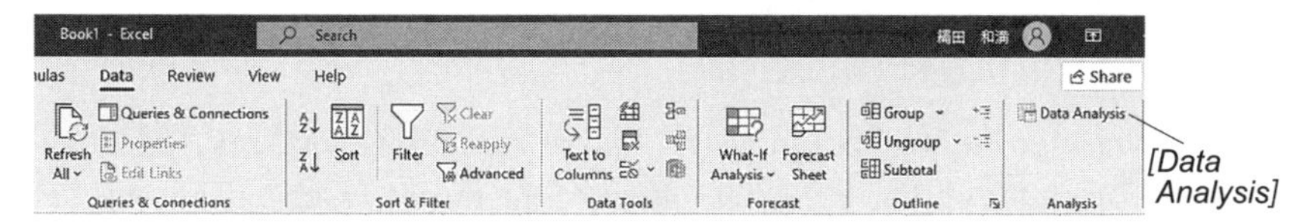

Fig. 1.10 Click the [Data] tab and check [Data Analysis].

[Add-in] on the left side of the box (Fig. 1.7). If [Options] does not appear, click [Move...]. The box changes to [Excel Options]. Check that [Excel Add-ins] is in [Manage](Fig. 1.8). Click [Go], and the [Add-ins] box appears. Then click the checkbox square in front of [Analysis ToolPak] so that it is checked, and click [OK] (Fig. 1.9). The [Analysis ToolPak] is built in. Click the [Data] tab and check [Data Analysis], which appears in the [Analysis] group (Fig. 1.10).

1.2 Input Data into Excel

Be sure that the [Home] tab is selected. (In this chapter, we do not use other tabs.) Excel is classified as spreadsheet software. We input the data per cell. A cell is identified by the combination of its column and row, which is called the "Cell Address." The columns are named using letters and the rows are named using numbers. For example, the first cell in the upper left corner is A1, and the cell in the third column and second row is C2. The cell in which we are inputting data is called the "Active Cell." We can move the active cell by:

i) Using the arrow keys (up, down, right, left; ↑, ↓, →, ←), and
ii) Moving the mouse pointer to the targeted cell, and clicking once.

For practice, try moving the active cell to D10.

We can input letters and other characters, numbers (numerical values), and equations and functions into the active cell.

1.2.1 Inputting Alphabets and Other Characters

Let A1 be the active cell. Type **New York** and press the [Enter] key (Fig. 1.11). As a result, "New York" is input into A1, and the active cell moves down by one cell so that A2 becomes the active cell. We can input up to 255 characters in a cell.

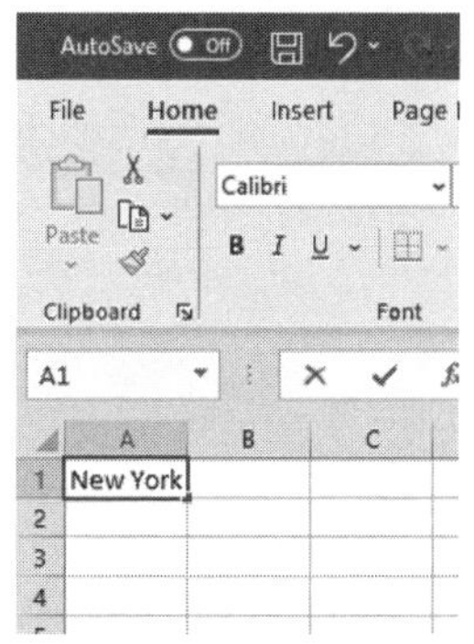

Fig. 1.11 Type **New York** in A1 and press the [Enter] key.

Let's input a sentence in A2. Be sure that the active cell is A2. Type **I go to London from**

NewYork, and press the Enter key. The sentence is input into A2, and the active cell becomes A3.

We can change the sentence in A2 to "I go to Paris from NewYork." Move the active cell back to A2. Click the left button of the mouse twice quickly. We call this procedure "double-clicking," and it is a widely used procedure in today's PCs. We can then change the contents of the cell. Using the [Arrow], [Delete], and [Back Space] keys, delete "London" and replace it with **Paris** and press the [Enter] key. The content of A2 is changed.

1.2.1 Changing Positions of Characters

In Excel, characters are presented aligned to the left; that is, the contents of the cell are aligned to the left. However, sometimes this alignment is improper, such as when we want to make a table. In such a case, we can change the position of the characters. So, let us change the positions of contents of cells. Let's make them align to the right. Move the mouse pointer to A1 and press the left button of the mouse. Hold the button down and move the mouse pointer to A2. The color of the cells changes and they are covered with bold lines (Fig. 1.12). Stop pushing the mouse button. This procedure is called "dragging," and it is an essential procedure for using Excel. Next, under the [Home] tab, choose the [Align Right] command button in the [Alignment] group and click it. As a result, the contents of the cells are presented as right-aligned. To left-align them again, choose the [Left Align] command button and click it. If we want to center the characters in the cell rather than align them to the right or left, click the [Center] button (Fig. 1.13).

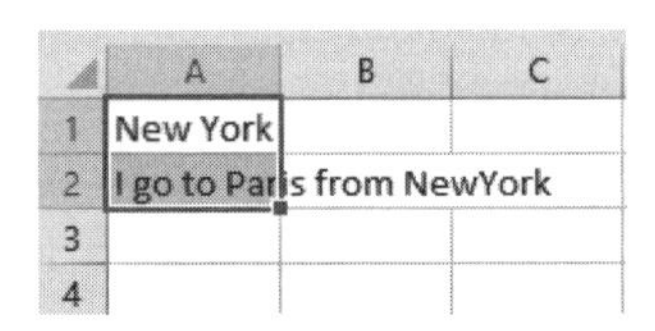

Fig. 1.12 Select the range from A1 to A2 by dragging. "Dragging" is an essential procedure for using Excel.

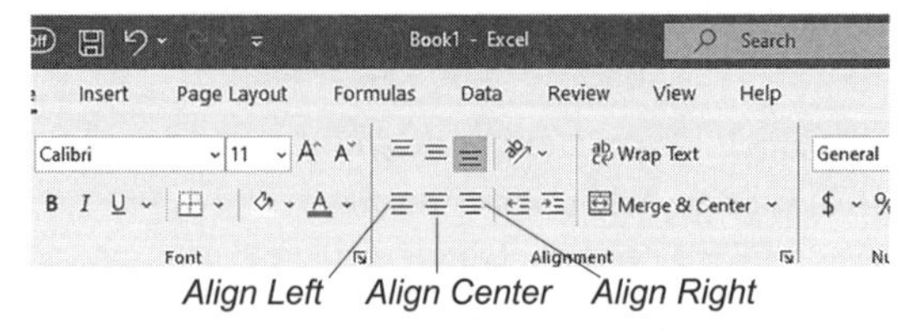

Fig. 1.13 Choose the [Align Right] command bottom in the [Alignment] group and click it. As a result, the contents of the cells are presented as right-aligned. To make them left-aligned, choose the [Left Align] button and click it. If we want to center the characters, click the [Center] button.

1.2.3 Inputting Numbers (Numerical Values)

Let's input numbers (numerical values). In cells A3 to A7, input the following numbers in order, one number per cell.

```
     123
   12345
   0.123
0.000001
 1000000
```

We can change the formats of these numbers to fixed point, comma, percent and scientific styles.

a. Change Digits after the Decimal Point

Let 123 be 123.0 with one digit after the decimal point. Make sure the active cell is A3, and choose [Increase Decimal] in the [Number] group and click it. Then, the number becomes 123.0. If we want to express the number up to two digits, as 123.00, click [Increase Decimal] again. To the contrary, if we want to reduce the number of expressed digits, click [Decrease Decimal] in the [Number] group (Fig 1.14).

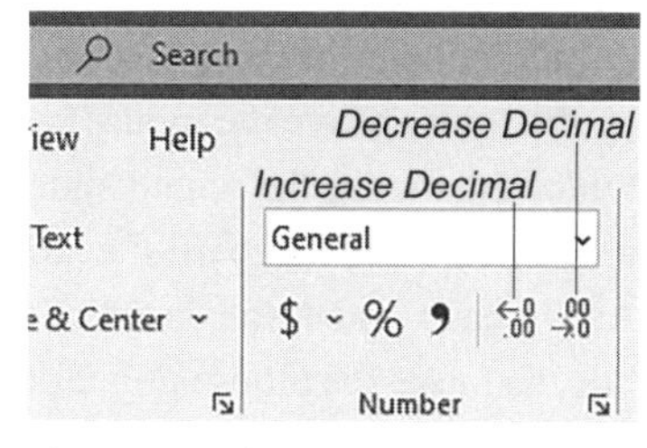

Fig. 1.14 Change digits below the decimal point.

b. Comma Style

Next, let the number 12345 be written in the comma style as 12,345. Move the active cell to A4. Choose the [Comma Style] button in the [Number] group and click it. Then the expression becomes 12,345. By combining the fixed point operations, the number can be expressed as 12,345.0. If we input a number as **123,456**, the format of that cell becomes comma style. When we change the number in that cell, the format of the cell remains comma style (Fig. 1.15).

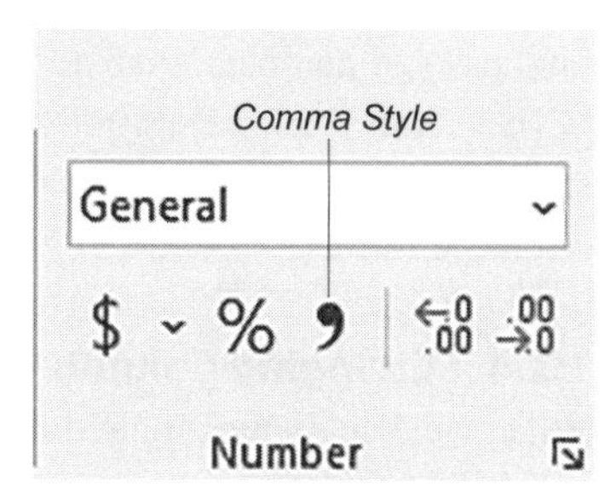

Fig. 1.15 Comma style.

c. Percent Style

Move the active cell to A5. Now, let 0.123 be 12.3%. Choose [%] (Percent Style) in the [Number] group and click it (Fig. 1.16).

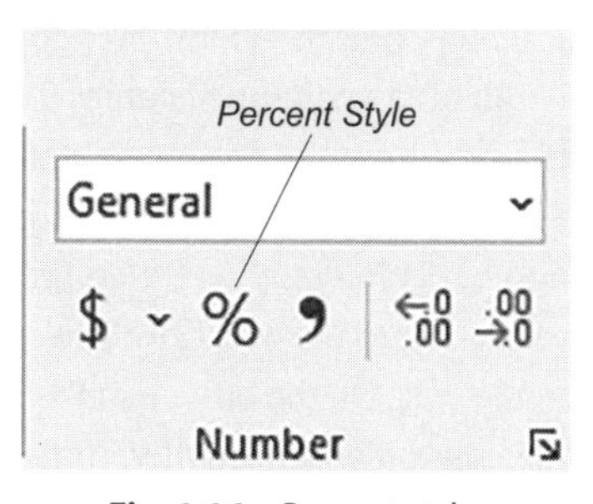

Fig. 1.16 Percent style.

Then, the expression becomes 12%. Next, choose [Increase Decimal] and click it. Then the expression becomes 12.3%. If we input **12.3%**, the format of the cell becomes "Percent Style" and is expressed as 12.30% with two digits after the decimal point. The digits after the decimal point can be changed as described in the description of the fixed point format.

d. Scientific Format (Exponent Notation)

When we use a very large (absolute) number such as 100000 or a very small number such as 0.000001, the scientific format or exponent notation is used. In this format,

1000000 becomes 1.00×10^6 and

0.000001 becomes 1.00×10^{-6}

In Excel they are expressed as 1.00E−06 and 1.00E+06. First, select A6 and A7 by dragging. Move the mouse and position the mouse pointer at A6. Hold the button of the mouse down and slide the mouse pointer to A7. Then, both A6 and A7 are selected and let go off the mouse button. The color of the two cells is changed. As we have noted, this is called "dragging." [Scientific Format] is not in the ribbon, so move the mouse pointer to the down arrow (∨) in the right side of the box [General]. Submenus appear (Fig. 1.17). Choose [Scientific]. (Hereafter, I will write "choose" or "select." Be sure to click the mouse button.) Then the expressions become 1.E−06 and 1.E+06. If we want to change the expression of the decimal parts of numbers, such as 1.000E−06 or 1.0000E06, use [Increase Decimal] or [Decrease Decimal] as in the fixed point format. We can input very large and small numbers as **1.0E6** and **1.0E−6**, and the format of the cell then becomes the scientific format.

1.2.4 Changing Column Width

When we display the inputted data, it is often necessary to change the width of a column. Click [Format] in the [Cells] tab. The submenu of [Format] appears, and we choose [Column Width] in the submenu (Fig. 1.18).

The [Column Width] box appears, so input **9** and click [OK] (Fig. 1.19). The width of the column becomes 9 characters. Input **123456789** in A8. Let this number be in comma style. As before, make A7 the active cell and click [Comma Style]. Then the cell becomes [########], and this means that the length of the number is longer than the column width and Excel cannot display the number correctly due to the overflow. Make the column width **15**. Then 123,456,789 appears in the cell. On the other hand, when we make a table, we sometimes want to make the column width smaller. In these cases, we can reduce the column width by the same method.

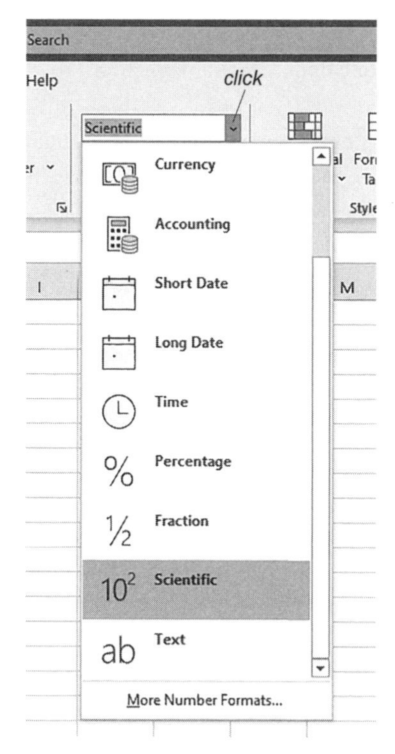

Fig. 1.17 Scientific style.

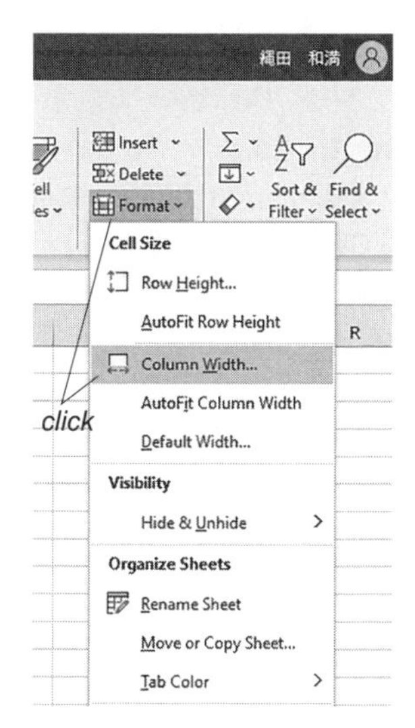

Fig. 1.18 To change the column width, click [Format] in the [Cells] tab. The [Format] submenu appears, and we choose [Column Width] in the submenu.

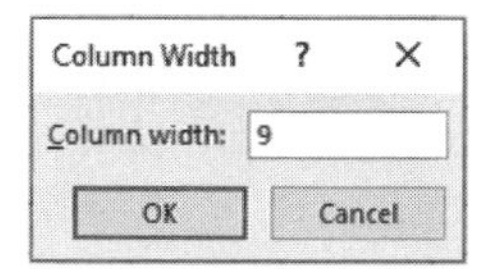

Fig. 1.19 The [Column Width] box appears, so input **9** and click [OK].

1.2.5 Inputting Formulas and Functions

a. Equations

In Excel, we can do various calculations very easily. Move the active cell down. Input **=2+3**, then we get 5, which is the result of the calculation. When we input equations, put = (or +, −) at the head. If we input **2+3**, it is interpreted as characters and 2+3 appears rather than the result of the calculation. The mathematical symbols used in Excel are: addition **+**, subtraction **−**. multiplication *****, division **/**, and exponentiation **^**. The order of calculation is exponentiation → multiplication and division → addition and subtraction in accordance with the usual mathematical rules. When we want to change the order, we use **( )**.

For example, $(2+3)^2/2$ is **=(2+3)^2/2** , $\{2+(3+2)^2\}\times 3$ is **=(2+(3+2)^2)*3**. (We do not use { } or [ ].) We get the results 12.5 and 81 in the inputted cells. Note that when we input **= −3^2**, Excel interprets it as the square of −3 and gives 9. Therefore, if we want to obtain the negative of 3 raised to the 2nd power, we need to input =−(3^2). However, **=1−3^2** is correctly calculated and we get −8.

b. Functions

Excel has many mathematical functions. For example, $\log_{10}(5)$ is calculated using **=LOG10(5)**.The names of functions can be both upper and lower case letters. We obtain the same results if we input either **=log10(5)** or **=Log10(5)**. As an argument in parentheses, we put a number or cell address that contains an objective number such as **=LOG10(A10)**. The functions can be used in the equations. $2\times\text{Log}_{10}(5)\times\text{Log}_{10}(7)$ is obtained by **=2*LOG10(5)*LOG10(7)**.

The major functions are as follows. Excel contains many other functions. For readers interested in the details, see the textbooks in (**Excel, Functions and Formulas**) of References.

ABS(number) Absolute value of the number

COS(number) Cosine of the number

EXP(number) Exponentiation of $e = 2.718....$

INT(number) Smallest integer that is larger than the number. If the number is positive, the integer part of the number. If the number is negative, the integer part of the number −1.

LN(number) Natural logarithm with base $e = 2.718....$ of the number

LOG(number1, number2) Logarithm of the number1 with a base of number2. If the number2 is omitted, it becomes a common logarithm with base 10.

LOG10(number) Common logarithm with base 10 of the number

MOD(number1, number2) Remainder of the number1 divided by the number2

PI() Circular constant, pi = 3.1415...

SIGN(number) Sign of the number

SIN(number) Sine of the number

SQRT(number) Square root of the number

RAND() Generation of a uniform random number between 0 and 1

ROUND(number1, digit) Round number1 off to the digit

c. Inputting Special Characters

Let inputting **1/2** signify the first page of a document consisting of two pages. Then the display of that cell becomes "2-Jan" or "1/2/2020" (it may change depending on the date). Excel treats 1/2 as date information. In this case, we need to input **'1/2**. Then, Excel displays 1/2 as the original input. This is the same as **1-2**, so try it. In some cases, it is necessary to treat numbers just as characters, and we use the same input method.

1.2.6 Inserting and Deleting Columns and Rows

After inputting the data, it often happens that we insert a new row or column. Insert a new row at the third row. Move the active cell to a cell in the third row, for example A3. Click the downward arrow [∨] on the right side of the [Insert] box in the [Cells] group. The submenu of [Insert] appears. Click [Insert Sheet Rows] (Fig. 1.20). Then a new row is inserted at the third row. We can insert a new column by the same method. When we want to delete an unnecessary row or column, move the active cell to the row or column we want to delete. Click the downward arrow [∨] on the right side of the [Delete] in the [Cells] group. Among the submenus, choose [Delete Sheet Rows] or [Delete Sheet Columns] (Fig. 1.21). Then a row or column is deleted. Be sure that there are no necessary data in the deleted rows or columns.

1.2.7 Saving and Opening a File

a. Saving a File the First Time

In Excel a unit of a work is called a "book". Let's save the file we are working on now. Select the [File] tab by moving the mouse pointer and clicking. (hereafter, I will write "select" or "choose" for this operation). Select [Save As] → [Browse] (Fig. 1.22). The [Save As] box appears. Select [Desktop]. Select the box on the right side of [File name:], and using the [Back Space] and [Delete] keys, delete the file name, Book1, in the box and type **EX1** (Fig. 1.23). In a file name, we can use up to 255 characters. However, since some special characters cannot be used, it is better for us to use 26 letters and the integers 0 ~ 9 until we become more familiar with Excel. In file names, upper and lower case characters are not

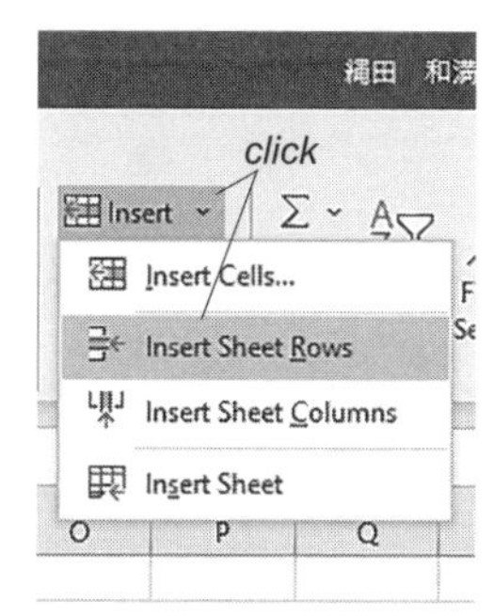

Fig. 1.20 Click the downward arrow [∨] on the right side of the [Insert] box in the [Cells] group. The [Insert] submenu appears, and click [Insert Sheet Rows].

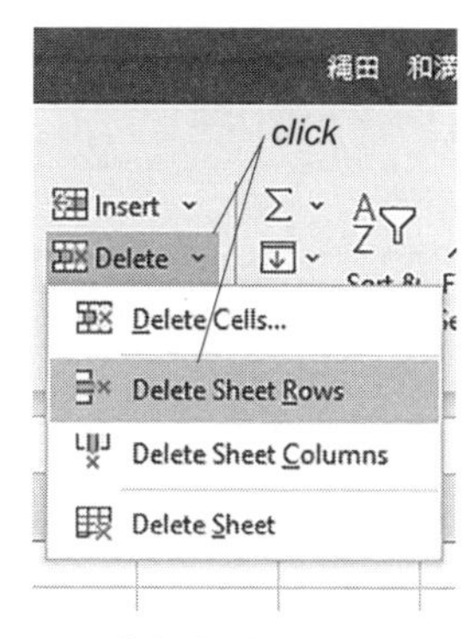

Fig. 1.21 Click the downward arrow [∨] to the right of [Delete]. In the submenu, choose [Delete Sheet Rows].

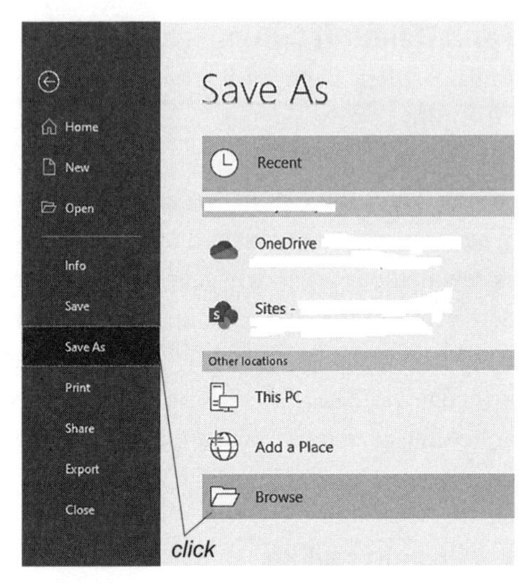

Fig. 1.22 Select [File] tab → [Save As] → [Browse].

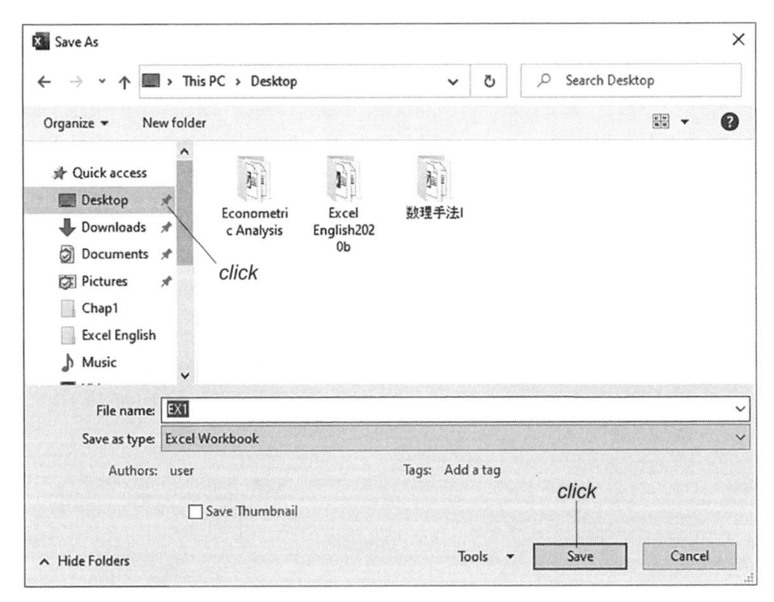

Fig. 1.23 Select [Desktop] and make the [File name:] **EX1**. After typing the file name, click [Save]. The workbook is saved as a file on the [Desktop], which is the startup screen.

distinguished, i.e., EX1, Ex1, ex1, and eX1 are the same file name in Windows. After typing the file name, click [Save]. The book is saved as a file on [Desktop], that is, on the startup screen.

b. Closing Excel

To finish our work in Excel, click the [Close] button [×] in the upper right corner (Fig. 1.24). If we do not save the workbook before we close Excel, all of the work will disappear. (If we try to close Excel without saving the workbook, Excel asks if we want to save the workbook or not.)

c. Opening a File

Click [Start] → [Excel] and start Excel again. We open the file made in previous sections. Choose [Open] in the menu. Choose [Browse] → [Desktop] (Fig. 1.25, 1.26). The [Open] box appears. Click [EX1] and [Open] (Fig 1.27). When we use the same PC continuously, we can open just by clicking the file name in the [Open] box (Fig. 1.26).

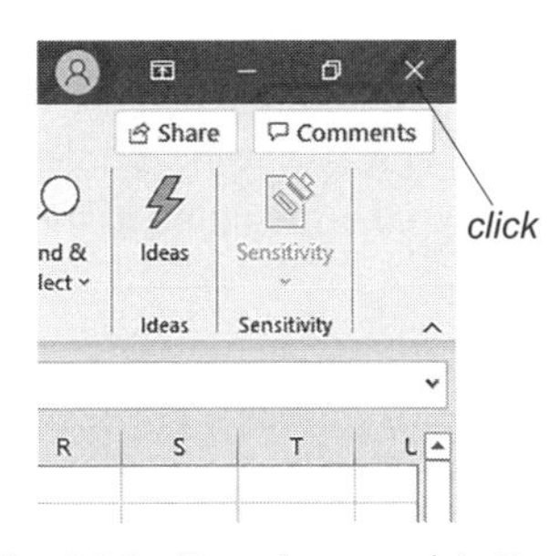

Fig. 1.24 To end our work in Excel, click the [Close] button [×] in the upper right corner.

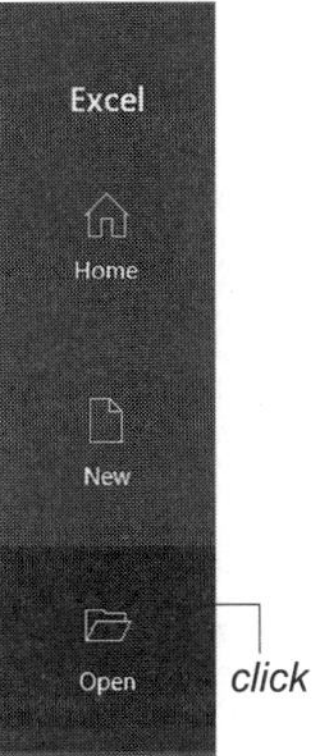

Fig. 1.25 Click [Start] → [Excel] and start Excel again. We will open the file created in the previous sections. Choose [Open] in the menu.

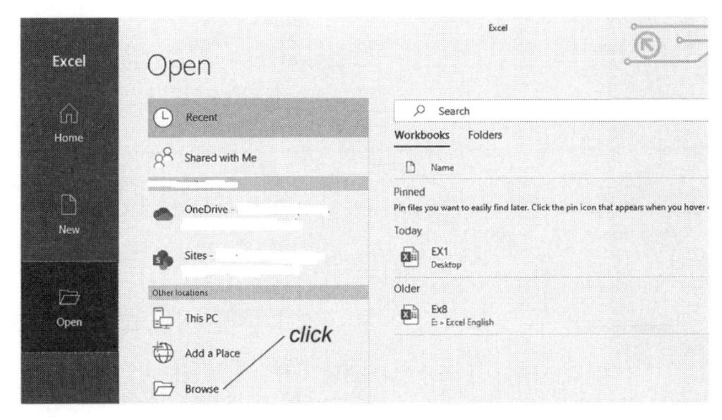

Fig. 1.26 Choose [Browse]. When we use the same PC, we can just click the file name in the [Open] box.

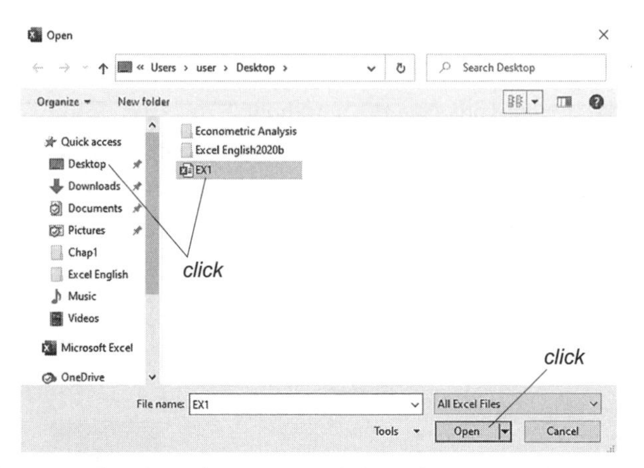

Fig. 1.27 The [Open] box appears. Click [Desktop] → [EX1] → [Open].

d. Saving the File after the First Time

When we save the file the second time and subsequently, choose [File] → [Save] (Fig. 1.28). Since the file is replaced with a new one, follow the procedures in (a) if you need both the new and old files, and change the file name of the new one to a related name such as **EX1B**.

e. Saving a File to Other Devices

In (a), we saved the file on the [Desktop] of the PC for simplicity. However, we

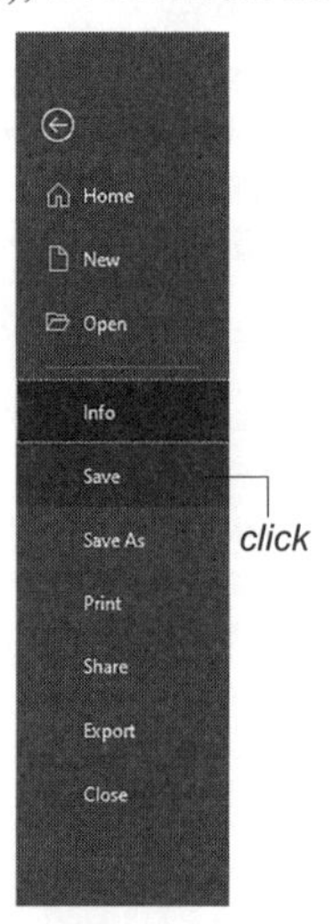

Fig. 1.28 When we save the file the second time and all subsequent times, choose [File] → [Save].

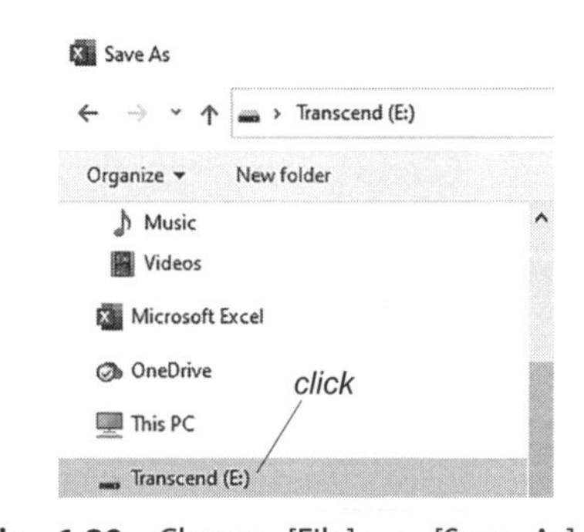

Fig. 1.29 Choose [File] → [Save As] → [Browse] as shown in Fig. 1.22. Then choose the target device, such as [USB DISK (E:)]. If the device does not appear, drag the scroll bar.

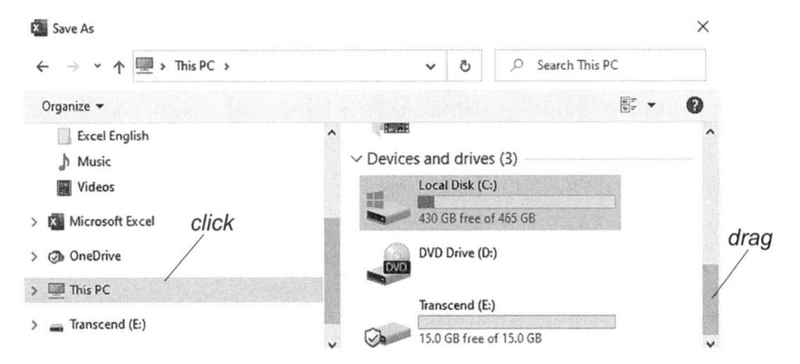

Fig. 1.30 If the target device is not displayed after performing the procedure shown in Fig. 1.29, click [This PC] and drag the scroll bar on the right side of the box.

often need to save a file on a different device such as a USB memory drive so that we can transport it easily. Let's save the file in a USB memory device. Attach the USB memory drive into the PC. Choose [File] → [Save As] → [Browse] (Fig. 1. 22). Then choose the target device such as [USB DISK (E:)] (Fig. 1.29). (E: is the device name assigned by the PC and it may differ depending on the PC). If the target device is not displayed when following the procedure shown in Fig. 1.29, click [This PC] and drag the scroll bar on the right side of the box. The devices we can use are displayed (Fig. 1.30).

1.3 New Folders, Passwords, and Cloud

1.3.1 Creating a New Folder

In previous sections, we saved the file on the [Desktop]. The advantage of doing so is that we can see the file immediately after starting the PC. However, the size

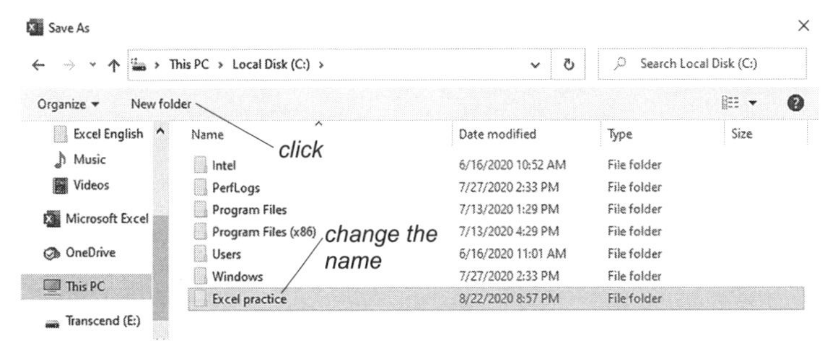

Fig. 1.31 Choose [File] → [Save As] → [Browse]. Click [This PC]. On the right side of the box, [Local Disk (C:)] appears in [Devices and drives] (Fig. 1.30, drag scroll bars until [This PC] and [Local Disk (C:)] appear.) Double click [Local Disk (C:)] and click [New folder]. A [New folder] is added. Then click it twice (but do not double click), and change the name of the new folder to **Excel practice**.

of the [Desktop] is limited, and we cannot save many files systematically in this way. We usually make folders and save files in the folders systematically. Let's make a folder. Choose [File] → [Save As] → [Browse]. Then, the [Save As] box appears. Click [This PC]. On the right side of the box, [Local Disk (C:)] appears in [Devices and drives] (drag the scroll bars until [Local Disk (C:)] appears). [C:] usually represents the hard disk of the PC (Fig. 1.30). Double click [Local Disk (C:)] and click [New folder]. A [New folder] is added. Then click it twice (do not double click), and change the name of the new folder to **Excel practice** (Fig.1.31). Now we can save and open files in the new folder by clicking [This PC] → [Local Disk (C:)] → [Excel practice].

1.3.2 Password and Encryption

If a file contains important information such as personal information, we can encrypt the file. Using the procedures already studied, open the [Save As] box so that we can save the file. Input a file name such as **EX1C**, and click [Tools] and choose [General Options] in the submenu (Fig. 1.32). The [General Options] box appears. Click the box on the right side of [Password to open] and type the password **Read1**, and also type the password **Write1** as the [Password to modify] (Fig. 1.33). In this case, upper and lower cases characters are treated as different characters, and we cannot open the file with **read1**. Click [OK]. The [Confirm Password] box opens, so reenter the correct passwords. Be aware that the file cannot be opened or modified if we forget the passwords (Fig. 1.34).

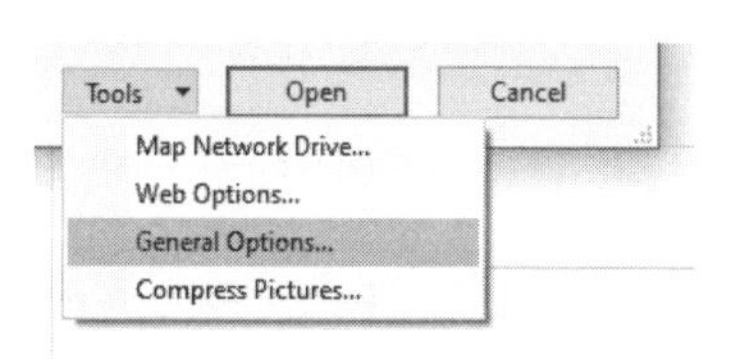

Fig. 1.32 Input the file name, and click [Tools] and choose [General Options] in the submenu.

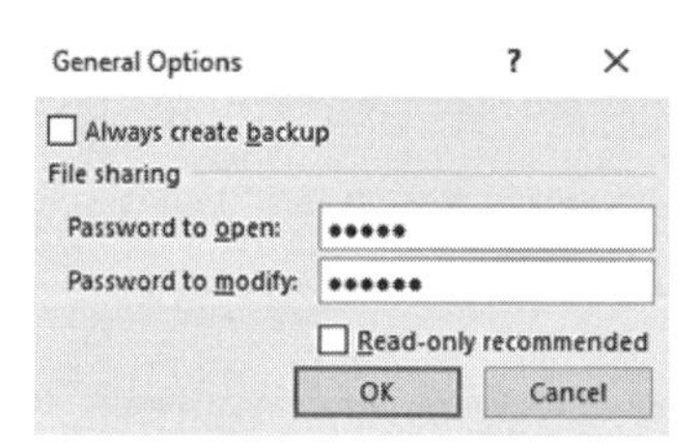

Fig. 1.33 The [General Options] box appears. Click the box on the right side of [Password to open] and type the password **Read1**. Also, type the password **Write1** as the [Password to modify]. In this case, upper and lower case characters are treated as different characters, and we cannot open the file by using **read1**. Click [OK].

Confirm Password
Reenter password to proceed.
Caution: If you lose or forget the password, it cannot be recovered. It is advisable to keep a list of passwords and their corresponding workbook and sheet names in a safe place. (Remember that passwords are case-sensitive.)
OK Cancel

Fig. 1.34 The [Confirm Password] box opens. Reenter the correct passwords. Be aware that the file cannot be opened or modified if we forget the passwords.

1.3.3 Cloud Service of Microsoft

When using multiple PCs in different places (for example, at an office or school and at home), it used to be common to use USB memory drives to transfer files between PCs. However, it was easy to forget or lose the USB drives. Now, we can use the Cloud service of Microsoft. I have been using the [One Drive] service. The files are stored in Microsoft's servers, and we can use the files throughout the world if the PC is connected to the internet. We need to have a Microsoft account, and the types of services and their prices vary. So, check the Microsoft web page or contact Microsoft directly for details.

1.4 Data Input Exercises

1. Make an Excel document in which you introduce yourself.
2. Table 1.1 shows the changes in the world population over time. (The data are from the Demographic Yearbook 2017 of the United Nations, and the units used are millions.) Input the data in Excel. By changing the alignments and widths of the cells, make the table easy to see. Save the file as **POP1**. We will use it in later exercises. Input the title of the table, **Table 1.1 World Population by Major Regions**, in A1. Input the years in Column A (put **Y** in front of each year so that Excel recognizes the years as characters), regions in Row 2, and population data in the cells from B3. Save the file as **POP1**.

Table 1.1 World Population by Major Regions

Year	World	Africa	Latin America	North America	Asia	Europe	Oceania
Y1960	3,033	285	221	205	1,700	606	16
Y1970	3,701	366	288	231	2,138	657	20
Y1980	4,458	480	364	254	2,642	694	23
Y1990	5,331	635	446	280	3,221	722	27
Y2000	6,145	818	526	313	3,730	727	31
Y2010	6,958	1,049	598	343	4,194	737	37
Y2017	7,550	1,256	646	361	4,504	742	41

Unit: millions; Source: Demographic Yearbook – 2017, United Nations

2

Spreadsheet Work in Excel

Excel is classified as spreadsheet software, and the spreadsheet application is one of its most important features. Using equations and functions, we can do complicated calculations and data processing easily. In this chapter, we learn how to work with spreadsheets in Excel.

2.1 Calculation of Sums of a Table

2.1.1 Inputting Equations

Start Excel as described in the previous chapter. Input the following numbers from A1 to B5

1	**6**
2	**7**
3	**8**
4	**9**
5	**10**

Let's calculate the sums of the rows and columns. First, calculate the sum of A1 and B1 in C1. Move the active cell to C1. Since we are calculating the sum of A1 and B1, the equation becomes A1+B1 (not 1+6). However, if we input **A1+B1**, Excel recognizes them as characters. So, adding =, input =**A1+B1** and push [Enter key] (Fig. 2.1). Then, 7 appears in C1. Note that cell addresses can be either lower or upper case letters. A1 and a1 are the same.

2.1.2 Copy/Paste and Relative Cell Reference

Let's calculate the sums of the columns from C2 to C5. If we input =**A2+B2**, it is meaningless to use Excel. In this case, we use the copy and paste features of Excel to

	A	B	C
1	1	6	=A1+B1
2	2	7	
3	3	8	
4	4	9	
5	5	10	

Fig. 2.1 Input =**A1+B1** in C1

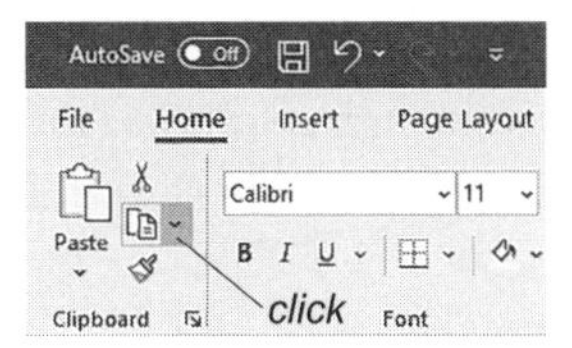

Fig. 2.2 Select [Copy] in the [Clipboard] group. (Move the mouse pointer to [Copy] and click it.) Then, the content of C1 is stored in the Windows [Clipboard].

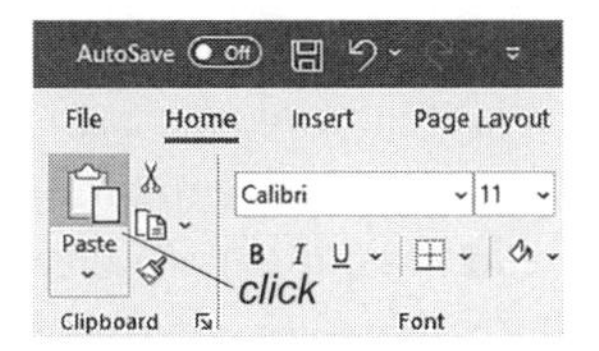

Fig. 2.3 Click [Paste] in the [Clipboard] group. The content of the clipboard is pasted, and the row sums are correctly calculated.

calculate the sums. Move the active cell to C1 and make sure that the tab is [Home]. Select [Copy] in the [Clipboard] group. (Move the mouse pointer to [Copy] and click it.) Then, the content of C1 is stored in the [Clipboard] of Windows (Fig. 2.2). Next, move the active cell to C2. Hold the left mouse button and move the mouse pointer to C5. After selecting C2 to C5, release the mouse button. The color of the selected cells (C2 to C5) changes. (This procedure is called "dragging".) Using the mouse, select [Paste] in the [Clipboard] group. Then, the content of the clipboard is pasted, and the row sums are correctly calculated (Fig. 2.3).

The content of C1 is =A1+B1; however, the content of C2, which was copied from C1, becomes =A2+C2 and we get the correct results. Why? The reason is that Excel uses "Relative Cell Reference." (Not only Excel but also all other spreadsheet programs use this method.) From C1, A1 is the second cell to the left and B1 is the first cell to the left. Inputting **=A1+B1** in C1 means "add the values of the second cell to the left and the first cell to the left." Therefore, by copying and pasting C1, C2 becomes =A2+B2, C3 becomes =A3+A3,..., and we can get the correct answers.

2.1.3 Calculation of Sums by Function

Next, let's calculate the sums of the columns. Move the active cell to A6. In this case, when we are calculating the sum of five cells, it is too much bother to input **=A1+A2+A3+A4+A5**. Excel has a function for calculating the sum of multiple cells. Input **=SUM(A1:A5)**. In the cell address, we can use both lower and upper case letters (i.e., sum(A1:A5) or Sum(A1:A5)) for functions. Copy A6 and paste it to B6 and C6 as before, and

SUM | =SUM(A1:A5)

	A	B	C	D	E
1	1	6	7		
2	2	7	9		
3	3	8	11		
4	4	9	13		
5	5	10	15		
6	=SUM(A1:A5)				

Fig. 2.4 We can easily calculate sums using the Excel function.

we can obtain the sums of the columns, i.e., 40 and 55, respectively (Fig. 2.4). The major functions that can be used for data processing are as follows (for readers who are interested in further details, see the textbooks listed in (**Excel, Functions and Formulas**) of References):

AVERAGE(range) Average of the cells in the range
CORREL(range1, range2) Correlation coefficient of range1 and range2
COUNT(range) Count the number of cells that have numerical values in the range
COUNTA(range) Count the number of non-empty cells in the range
COVAR(range1, range2) Covariance of range1 and range2
MAX(range) Maximum value of the range
MIN(range) Minimum value of the range
QUARTILE(range, integer) Quartile value determined by the integer in the range
STDEV(range) Sample standard deviation in the range
STDEVP(range) Population standard deviation in the range
VAR(range) Sample variance in the range
VARP(range) Population variance in the range

2.2 Calculation of Ratios and Absolute Cell Address

2.2.1 Ratios with Respect to the Total Sum

Let's calculate the ratios of the cell values to the total sum of 55. Move the active cell to A11. We input the cell addresses not just by typing but also by using the mouse. Just type = in cell A11. Move the mouse pointer to A1 and click it. Then A1 appears in A11. Type **/** and A11 becomes **=A1/**. Move the mouse pointer to C6 and click it. **=A1/C6** is inputted in A11. Press the [Enter] key, and the portion of A1 dividing by C6 (0.01818…) appears in A11. The method of inputting equations using the mouse is convenient when the table becomes larger.

We can use this method for functions. Let's calculate the sum of 10 cells, A1 to B5. In some cell, input **=SUM(**and move the mouse pointer to A1. Then drag A1 to B5 (holding the left button of the mouse and moving the mouse pointer to B5), and type). **=SUM(A1:B5)** is inputted in the cell and we get 55 when we push the [Enter] key.

Next, we calculate the portions of 6 rows × 3 columns. Excel uses the relative cell reference method, and so we cannot use the Copy/Paste method studied in the previous section. If we copy and paste A11, division by zero happens in the other cells, and errors (#DIV/0!) occur.

In this case, we use "Absolute Cell Reference." The absolute cell address fixes the cell address. Double click A11, and change C6 to **$C$6**, putting **$** in front of the letter representing a column and the integer representing a row. The cell address is fixed. $C$6 always refers to the contents of C6 and it does not change when we copy and paste.

After changing A11 to **=A1/$C$6**, push the [Enter] key. The displayed number is not changed. Let's copy and paste A11. First, move the active cell to A11 and select the [Copy] button as before. Drag from A11 to C16. We can copy and paste A11 to A11 itself; we do not have to drag the range twice or more. Select [Paste] in the Clipboard group. Since the numerators change but the denominators do not, we get the correct results in the table. Change the format of the cells to the percentage format with a digit below the decimal point to make the table easy to understand.

2.2.2 Ratios to Sums of Rows

Let's calculate the ratios of the cell values to the sums of the rows. Input **=A1/C1** in F1; that is the ratio of A1 to C1, the sum of the row cells. If we copy and paste F1 we cannot get the correct answers as in the previous case. In the ratios to the row sums, the denominators become C2, C3,... after the second cell. The row numbers change, but the column letter does not change. In this case, we combine the relative and absolute cell addresses and type **=A1/$C1**. The column is fixed, but the row number changes according to the position of the cell.

Copy F1, drag the range F1:H6, and select the [Paste]. Change the format of the cells to the percentage style with a digit below the decimal point to make the table

	A	B	C	D	E	F	G	H
1	1	6	7			14.3%	85.7%	100.0%
2	2	7	9			22.2%	77.8%	100.0%
3	3	8	11		=A1/$C1	27.3%	72.7%	100.0%
4	4	9	13			30.8%	69.2%	100.0%
5	5	10	15			33.3%	66.7%	100.0%
6	15	40	55			27.3%	72.7%	100.0%
7								
8								
9		=A1/$C$6					=A1/A$6	
10								
11	1.8%	10.9%	12.7%			6.7%	15.0%	12.7%
12	3.6%	12.7%	16.4%			13.3%	17.5%	16.4%
13	5.5%	14.5%	20.0%			20.0%	20.0%	20.0%
14	7.3%	16.4%	23.6%			26.7%	22.5%	23.6%
15	9.1%	18.2%	27.3%			33.3%	25.0%	27.3%
16	27.3%	72.7%	100.0%			100.0%	100.0%	100.0%

Fig. 2.5 Input the equations using the relative and absolute cell references and copy and paste to the tables.

easy to understand, as described above (Fig. 2.5).

2.2.3 Ratios to Sums of Columns

Next, let's calculate the ratios of the cell values to the sums of the columns. In F11, we will calculate the ratio of A1 to A6, the sum of that column. In this case, the row integer is fixed and the column letter changes. Input **=A1/A$6**, so that the row is fixed and the column changes, and copy F11 to the range F11:H16 as before. Change the format of cells to the percentage format with a digit below the decimal point to make the table easy to understand.

2.3 Drawing Ruled Lines

We have made four tables. Let's draw ruled lines so that we can read them more easily. Draw ruled lines for the table in the range A1:A6. Drag the range. Next, click the down arrow [∨] to the right of [Borders] in the [Font] tab (Fig. 2.6). The submenu appears. Click [All Borders] by the mouse button, and ruled lines appear in the selected region. If we want to delete the ruled lines, select [No Border] in the submenu of [Borders] (Fig 2.7). We can set the line style, such as to bold, dotted, or broken lines. Click [Line Style] in the submenu of [Borders]. The line submenu appears. Select a proper line style (Fig. 2.8). Draw ruled lines in the other three tables.

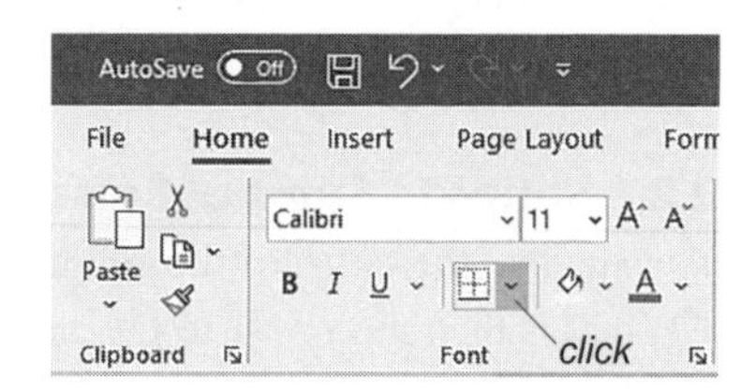

Fig. 2.6 For drawing ruled lines, select the table range and click the down arrow [∨] to the right of [Borders] in the [Font] tab.

2.4 Copying the Values of Cells

Now copy and paste the ratio tables to the other part of the worksheet. Since the relative cell references are used, we have to use the "paste values." Drag the range A11:A16 using the mouse, and click [Copy]. The contents of A11:A16 are stored in the clipboard. Then move to a cell we want to paste in the table, for example A21. We specify one cell at the beginning of the range and move the active cell to A21. (To copy a table of 6 rows and 3 columns, we just specify one cell. When we select multiple cells, errors may occur.) Click the down arrow [∨] below [Paste] (Fig. 2.9). The submenu of [Paste] appears. Click [Paste Special] at the bottom (Fig. 2.10). The box for [Paste Special] appears. Select [Values] and click [OK]

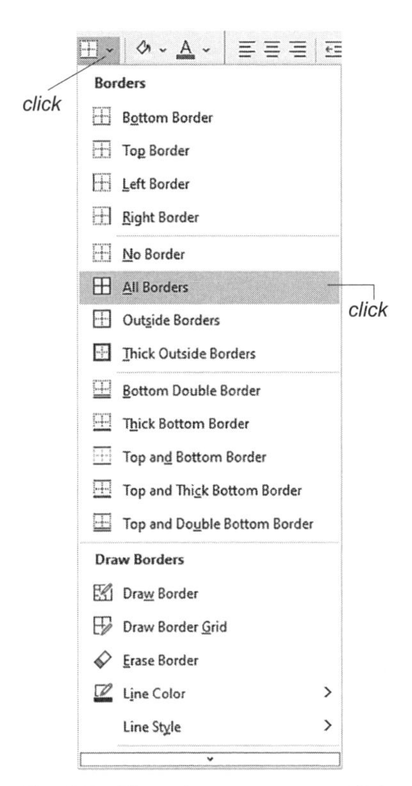

Fig. 2.7 The submenu appears. Click [All Borders] using the mouse button, then the ruled lines in the selected region. If we want to delete the ruled lines, select [No Border] in the submenu of [Borders].

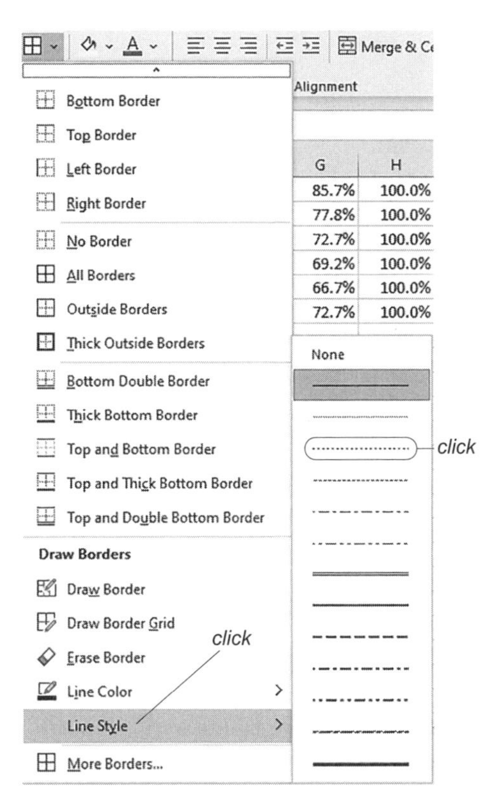

Fig. 2.8 We can choose the line style, such as bold, dotted, or broken lines. Click [Line Style] in the submenu of [Borders]. The line submenu appears. Select a proper line style.

(Fig. 2.11). Then the values (not equations) of A11:A16 are pasted to the range A21:A26. With this procedure, only the values are pasted. If we want to copy the formats of the cells (for example, the ruled lines and styles), click the down arrow [$\vee$] below [Paste] → [Paste Special] as before. This time, select [Formats] in the [Paste Special] box and click [OK]. Then the formats of the cells are copied. Copy the other tables to the proper ranges of the worksheet.

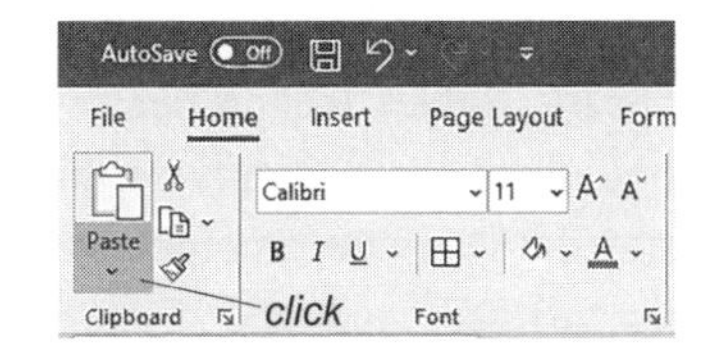

Fig. 2.9 Click the down arrow [$\vee$] below [Paste].

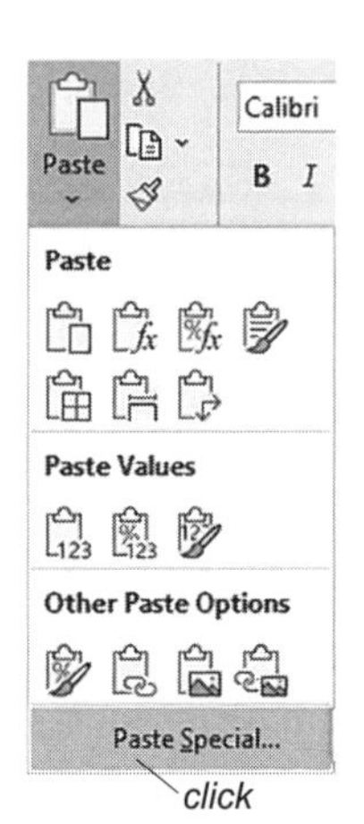

Fig. 2.10 In the submenu of [Paste], click the [Paste Special] button.

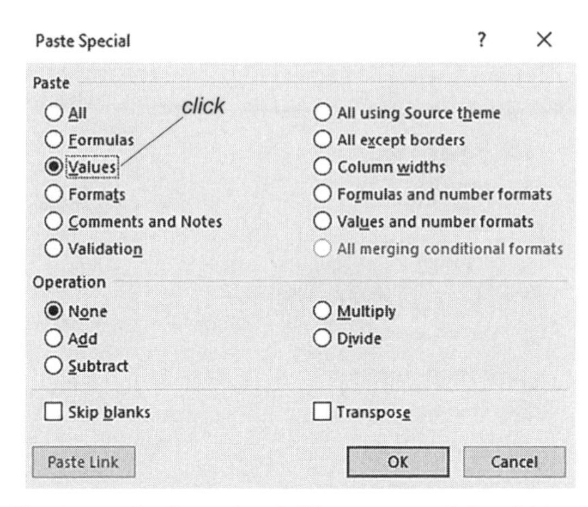

Fig. 2.11 The [Paste Special] box appears. Select [Values] and click [OK]. Then the values (not equations) of A11:A16 are pasted in the range of cells from A21 to A26.

2.5 Printing the Selected Range

Let's print out the tables we have made. Check that the printer is ready to print. Next, select the ranges we want to print (for example, the range A11:A16, the ratios to the total sum) by dragging. Select the [File] tab and click [Print] in the submenu (Fig. 2.12). Check whether the printer is selected correctly. If not, click the down arrow [∨] to the right of [Printer]. The list of printers appears. Select the proper printer. If we print now, all of the contents of the active sheet are printed. (Many people, including me, have made this mistake, and ended up with meaningless printed output.) So, click the down arrow [∨] to the right of [Print Active Sheets] and select [Print Selection]. Click [Print]. The range we have selected is printed. If we want to stop printing, click the [Back] button (Fig. 2.13).

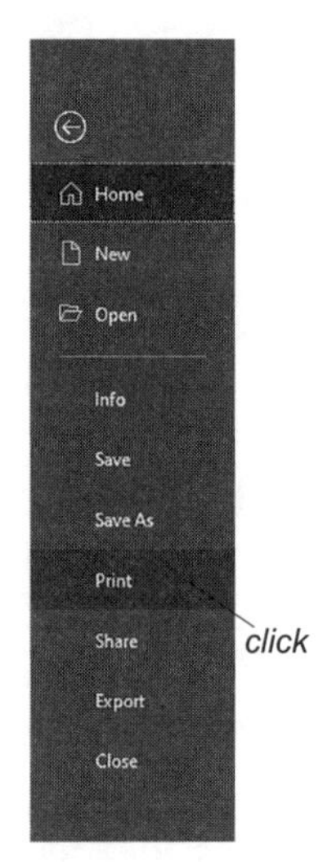

Fig. 2.12 Select the range of cells we want to print out by dragging the mouse pointer. Select the [File] tab and click [Print] in the submenu.

We can change the styles of the pages. Go back to the worksheet. Select the [Page Layout] tab.

The ribbon changes to that of [Page Layout]. If we want to change the margins or the header/footer of the page, click the down arrow [∨] at the bottom of [Margins] and choose [Custom Margins] (Fig. 2.14, Fig. 2.15). Then the [Page Setup] box appears. If we want to change the margins, choose [Margins] and change

Fig. 2.13 Check whether the printer is selected correctly. Click the down arrow [∨] to the right of [Print Active Sheets] and select [Print Selection]. Click [Print]. The range we have selected is printed. If we want to stop the printing, click the [Back] button.

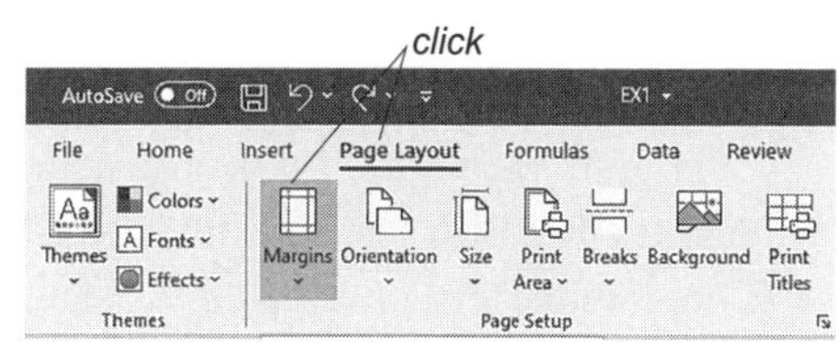

Fig. 2.14 Select the [Page Layout] tab. If we want to change the margins of the page, click the down arrow [∨] below [Margins].

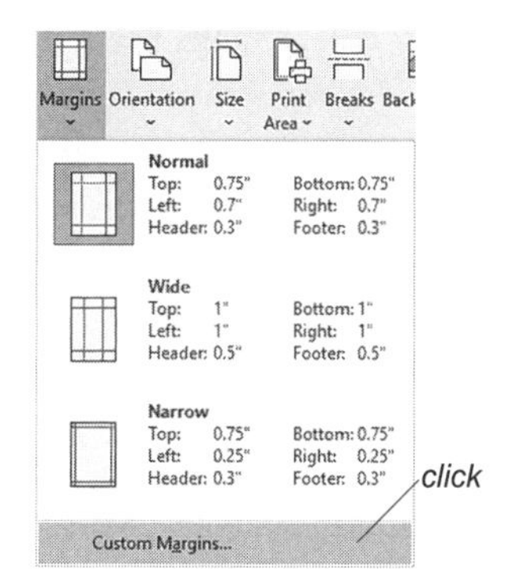

Fig. 2.15 Choose [Custom Margins].

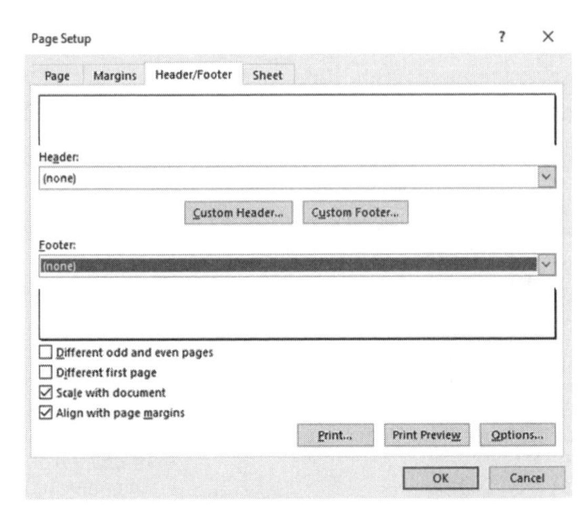

Fig. 2.16 The [Page Setup] box appears. If we want to change the header and footer, click [Header/Footer] and change them. Click [OK] when finished changing the page styles.

them to the proper ones. When we change the header and footer, click [Header/Footer] and change them. Click [OK] when we finish changing the page styles (Fig. 2.16). Then print the selected range as before.

In this chapter, we have studied basic spreadsheet features. Save the file as **EX2**.

2.6 Exercises Calculating Portions of Regional Population and Population Growth Rates

2.6.1 Portions of Regional Population

Here, we calculate the percentages of regional populations and the annual growth rates of the population. First select [File] → [Close] and close the current workbook. Do not forget to save the file. Open the file containing the world population that we made in the exercise in Chapter 1.

From K1, let's calculate the percentages of regional populations. Input **Table 2 Regional Population Rates** in K1. Copy and paste (hereafter, I will just write "copy." Do not forget to click [Paste]) the Years to the range starting with K2 (K2 represents the year). (Drag A2:A9. Click [Copy], move the active cell to K2, and click [paste].) Copy the names of the regions "World" to "Oceania" to range from L2 to R2.

The percentages of the regional populations are obtained by the Regional Population/World Population, and the denominators representing the world population are unchanged for the same year. So, combining the relative and absolute cell references, input **=B3/$B3** in L3 and copy and paste it to the entire table. We can then calculate the regional population percentages. Change the style of the cells to percentages with one digit under the decimal point and use ruled lines to make the table easy to read.

2.6.2 Annual Population Growth Rates

Next, we calculate the annual growth rates of the population. Let P_0 be the population in the base year, P_t be the population in year t, and r be the annual growth rate. Then, we get

(2.1) $$P_t = P_0(1+r)^t$$

Therefore, the annual growth rate is obtained by

(2.2) $$r = (P_t/P_0)^{1/t} - 1$$

Let's calculate the population growth rates in the range of cells beginning with A21. Input **Table 3 Annual Growth Rates of Regional Populations**. Input **Year** in A22 and copy regional names "World" to "Oceania" from B22 to H22. We calculate the growth rates from 1960 to 2017. Input **1960** in A23. In the same way, input **1970**, **1980**, **1990**, **2000**, **2010**, and **2017** from A24 to A29 as numbers. Using Equation (2.2), we calculate the annual population growth rates. For all regions, t is constant if the years are determined, and we combine the relative and absolute cell references. Input **=(B4/B3)^(1/($A24–$A23))–1** in B23 and copy it throughout the table (up to Row 28, not Row 29).

Numbers from 1960 to 2017 are input in Column A. We need to rewrite them as 1960-1970, 1970-1980,..., 2010-2017. However, 1960,..., 2017 are used in calculations, So, copy the table by copying and pasting the values in the cells to the proper place in the worksheet and replace 1960 with 1960-1970. Complete the table and make it easy to read by changing the cell styles (percentages with two digits after the decimal point) and using ruled lines. If you mistakenly copy unnecessary cells, delete them by pressing the [Delete] key.

We have reached the end of this section. Do not forget to save the file.

3

Making Graphs

We can easily make various graphs with Excel. In this chapter, we learn the methods for making graphs.

3.1 Bar Chart

3.1.1 Making a Simple Bar Chart

Let's start Excel and input the following data from A1:B6.

Year	Personal Consumption Expenditures
Y2000	6,977
Y2005	8,470
Y2010	10,002
Y2015	12,040
Y2019	14,228

The data are personal consumption expenditures in the United States of America (USA, seasonally adjusted annual rate in billions of dollars as of January of that year, obtained from FRED Graph Observations, Economic Research Division, Federal Reserve Bank of St. Louis). Put "**Y**" in front of the years so that Excel recognizes them as characters. Let's make a bar graph where the horizontal axis (X-axis) is the year. Using the mouse, drag A1:B6 and select the range. Then click the [Insert] tab. The ribbon changes to that of [Insert]. Choose [Insert Column or Bar Chart] in the [Charts] group (Fig. 3.1). The submenu of [Insert Column or Bar Chart] appears. Choose [Clustered Column] in [2-D Column] (Fig. 3.2). The bar chart appears and the ribbon is changed to the graph edition version. [Chart Tools] contains the [Design] and [Format] tabs. [Design] is selected. [Chart Tools] is not usually displayed, but it appears when we edit the graph. The graph is surrounded lines containing 8 dots (◦) and enters active status so that it can be edited.

Excel automatically determines the position of the graph. When the position of the graph is not correct, move it by clicking a part where nothing is displayed and

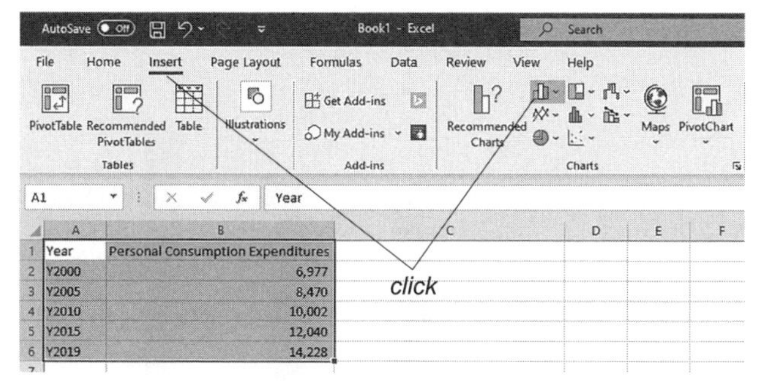

Fig. 3.1 Select the data range by dragging. Click the [Insert] tab. The ribbon changes to that of [Insert]. Choose [Insert Column or Bar Chart] in the [Charts] group.

dragging. (If we drag a part that has content, that part will move.) Since the graph is too small, let's make it larger. Move the mouse pointer to the dot at the lower right of the graph. Then, the arrow changes to a chart arrow (diagonal double-headed arrow). We can change the graph size by dragging the chart arrow (Fig. 3.3). Next change the sizes of the parts in the graph. Click the part we want to change so that the part is surrounded lines containing dots. Drag the dots at the corners and change the size of the selected part (Fig. 3.4).

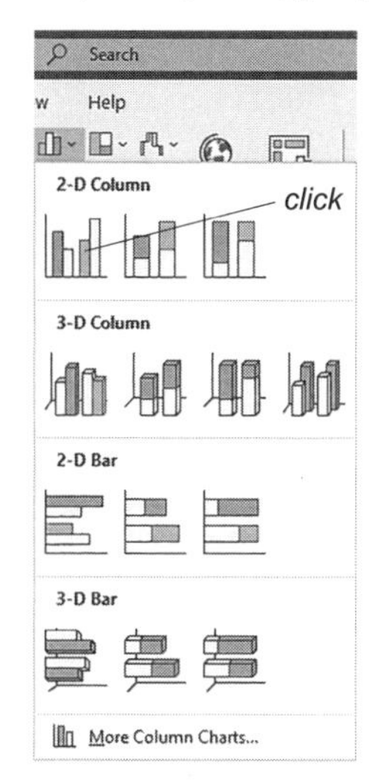

Fig. 3.2 The submenu of [Insert Column or Bar Chart] appears. Choose [Clustered Column] in [2-D Column].

Next, let's change the title of the graph and input axis labels using the following procedures. Make sure the graph is active so that it is surrounded by lines containing dots. If not, click the graph.

i) Move the mouse pointer to the graph title "Personal Consumption Expenditures" and click it so that the title is surrounded lines containing four dots at the corners. Then, we can edit the title and change it to **Fig. 1 Personal Consumption Expenditures** (Fig. 3.5).

ii) Input a horizontal axis label. Click the [Format] tab and change the ribbon, and click [Text Box] in [Insert Shapes] (Fig. 3.6). Move the mouse pointer (the shape of the pointer changes) to the bottom part of the graph where we want to input the label, and then determine the size of the

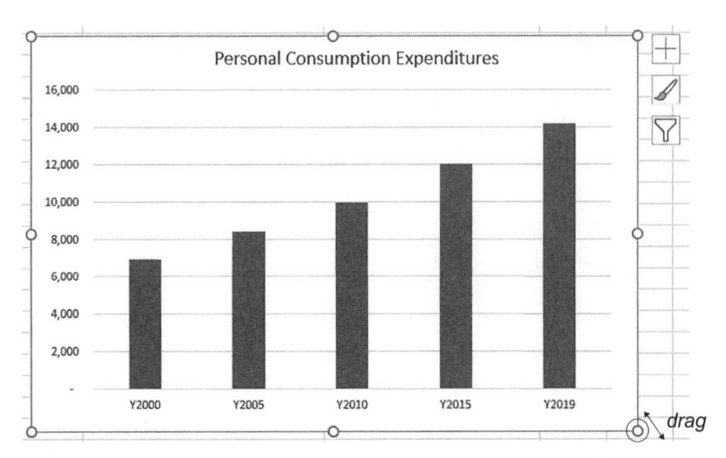

Fig. 3.3 When the position of the graph is not correct, move it by clicking a part where nothing is displayed and dragging. (If we click a part with content, that part will move.) Since the graph is too small, let's make it larger. Move the mouse pointer to the dot at the lower right of the graph. Then, the arrow changes to a chart arrow (diagonal double-headed arrow). We can change the graph size by dragging a chart arrow.

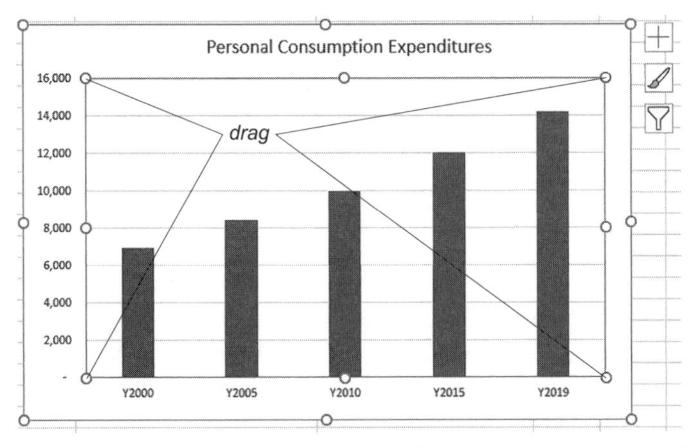

Fig. 3.4 To change the sizes of the parts in the graph, click the part we want to change so that the part is surrounded by lines containing dots. Drag the dots at the corners and change the size of the part.

label by dragging the mouse pointer, and input the label **Year** (Fig. 3.7).

iii) In the same way, input **Billion Dollars** as the vertical axis label (Fig. 3.8).

The font sizes of the title and axis labels may be too small or too large. If that is the case, we can change them. Move the mouse pointer to the title of the graph and click it so that it is surrounded by lines containing dots. Click the [Home] tab

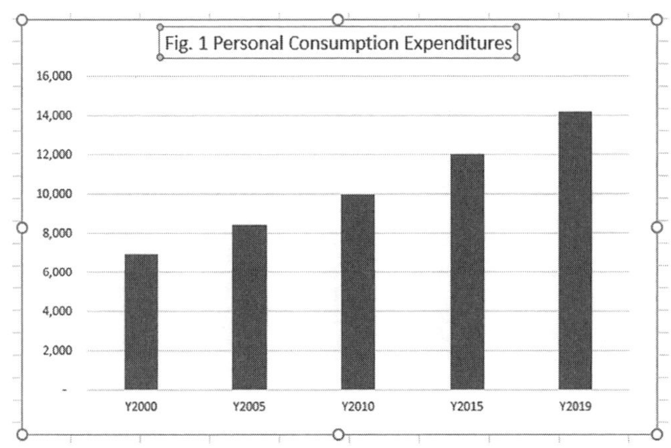

Fig. 3.5 Move the mouse pointer to the graph title "Personal Consumption Expenditures" and click it so that the title is surrounded by lines containing four dots at the corners. Then, we can edit the title and change it to **Fig. 1 Personal Consumption Expenditures.**

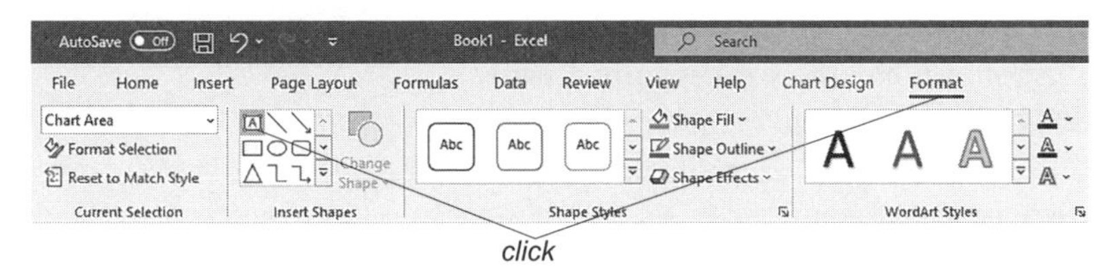

Fig. 3.6 Click the [Format] tab and change the ribbon, then click [Text Box] in [Insert Shapes].

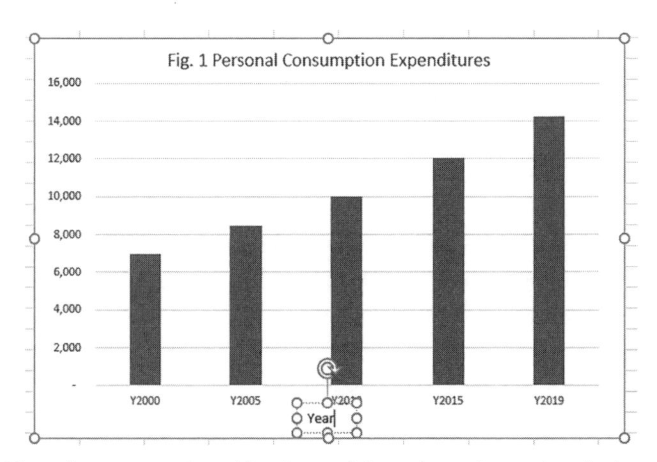

Fig. 3.7 Move the mouse pointer (the shape of the pointer changes) to the bottom part of the graph where we want to input the label, and determine the size of the label by dragging. Input the label **Year**.

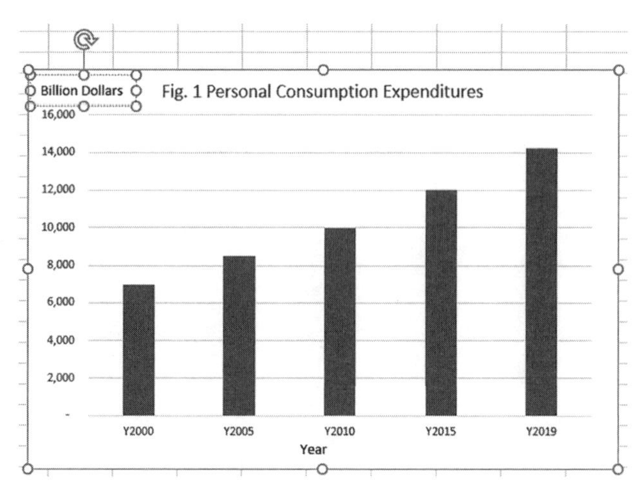

Fig. 3.8 Input **Billion Dollars** as the vertical axis label.

→ the downward arrow of the [Font Size] box. The list of possible font sizes appears. Change it to **[12]**. The characters become larger as the font size is increased. Change the font sizes of the axis labels so that we can read the graph easily (Fig. 3.9). After making the graph, click a cell outside of the graph area. Excel returns to the normal input mode. If we want to edit the graph again, click the graph and make it active.

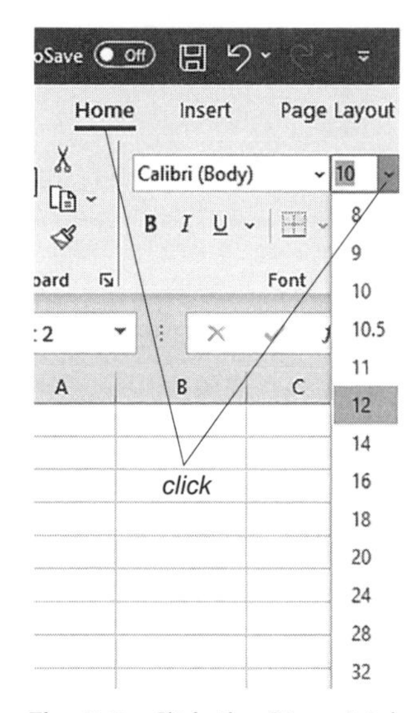

Fig. 3.9 Click the [Home] tab → the down arrow of the [Font Size] box. The list of possible font sizes appears. Change it to [12].

3.1.2 Printing the Graph

a. Printing with Excel

Let's print the graph made in the previous section. Click the graph and make it active. Select [File] → [Print]. The [Print] box appears. Check that the correct printer is selected. (If not, click the down arrow (∨) to the right of the [Printer] box and select the correct printer.) Click [Print], and the active graph is printed (Fig. 3.10).

b. Pasting the Graph to Word

In the previous section, we printed the graph. However, we often need to use graphs and tables as part of a report. Since Excel is not a word processor, it is bet-

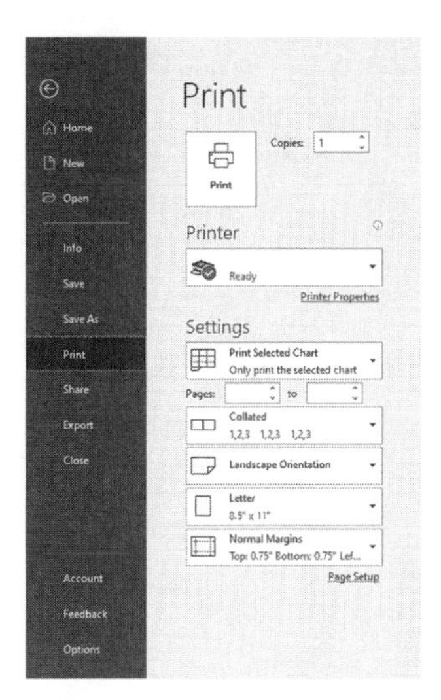

Fig. 3.10 To print the graph, click the graph and make it active. Select [File] → [Print]. The [Print] box appears. Check that the correct printer is selected. Click [Print], and the active graph is printed.

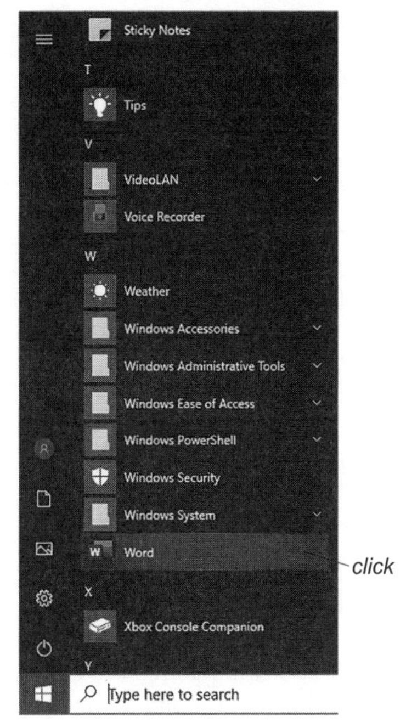

Fig. 3.11 Select [Start] → [Word] to start Word.

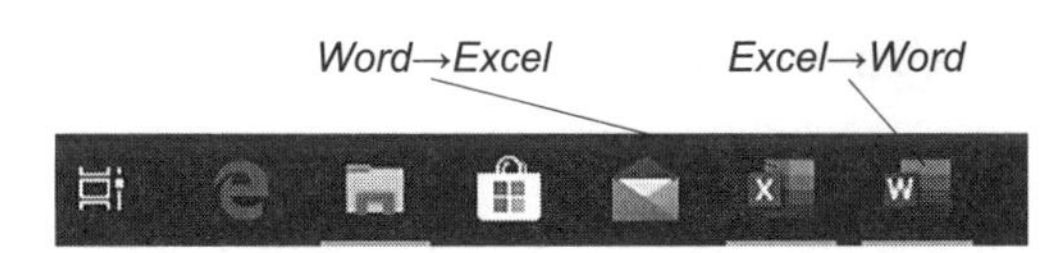

Fig. 3.12 We can switch between Excel and [Word] by clicking the [Excel] and [Word] icons at the bottom of the screen.

ter to use Word when making a report. Start Word by selecting [Start] → [Word] (Fig. 3.11, keeping Excel open).

Let's paste the graph into Word. Click the [Excel] icon at the bottom of the screen. The display is changed to Excel. Select the graph, check the [Home] tab, and click [Copy]. Then, click the [Word] icon at the bottom of the screen. The display is changed to Word (Fig. 3.12). Click the [Paste] button as in Excel. Then, the graph is pasted into Word. We can use [copy/paste] in both Excel and Word as if they were the same program. We can change the size of the graph by dragging the right lower corner of the graph in Word. Go back to Excel and select A1:B6 and copy it into Word using the same procedure (Fig. 3.13). You can print and save the file by the same methods used in Excel.

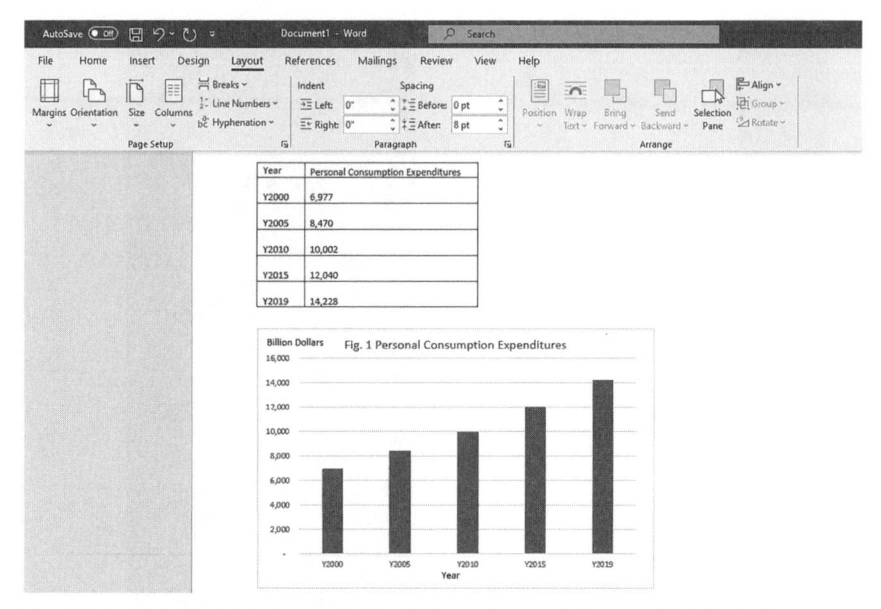

Fig. 3.13 Click [Paste] button as in Excel. Then, the graph is pasted in Word. We can use [Copy/Paste] in both Excel and Word as if they were the same program.

3.1.3 Bar Chart Using Multiple Variables

Next, we make a graph containing multiple variables. Input the following data into Column C (C1:C6).

Government Total Expenditures

3,076
4,196
5,820
4,196
7,150

The new data are the total government expenditures in the USA (seasonally adjusted annual rate as of the first quarter beginning in January, in billions of dollars, obtained from FRED Graph Observations, Economic Research Division, Federal Reserve Bank of St. Louis). Let's make a bar chart of "Personal Consumption Expenditures" and "Government Total Expenditures." Select the data range, A1:C6, by dragging. Click [Insert] → [Insert Column or Bar Chart] → [Clustered Column] under 2-D Column (Fig. 3.14, 3.15). A bar chart consisting of the two variables appears. Title the graph **Fig. 2 Personal Consumption Expenditures and Government Total Expenditures**, insert the horizontal and vertical

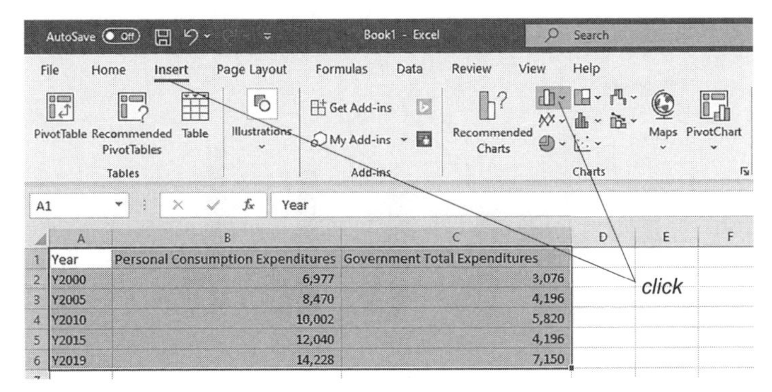

	A	B	C
1	Year	Personal Consumption Expenditures	Government Total Expenditures
2	Y2000	6,977	3,076
3	Y2005	8,470	4,196
4	Y2010	10,002	5,820
5	Y2015	12,040	4,196
6	Y2019	14,228	7,150

Fig. 3.14 Select the data range, A1:C6, by dragging. Click [Insert] → [Insert Column or Bar Chart]

axis labels, and complete the graph (Fig. 3.16).

Next, we make a stacked column bar chart. As before, select the range A1:C6 by dragging. Click [Insert] → [Insert Column or Bar Chart] → [Stacked Column] under 2D Column. (Fig. 3.17). Then the stacked column bar chart appears. Title the graph **Fig. 3 Personal Consumption Expenditures and Government Total Expenditures (Stacked Column)**, insert the horizontal and vertical axis labels, and complete the graph (Fig. 3.18).

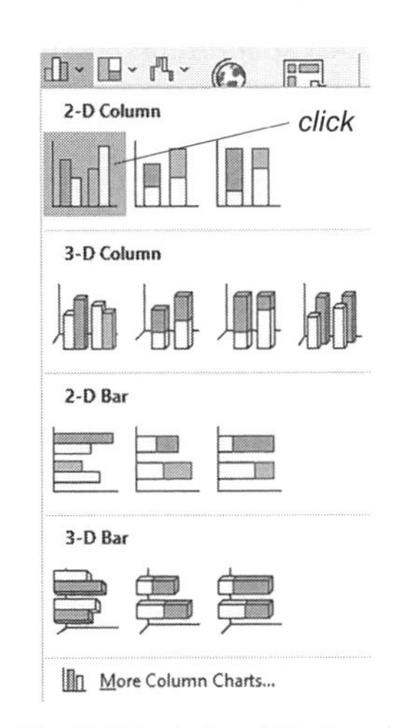

Fig. 3.15 Select [Clustered Column] under 2-D Column.

3.1.4 Bar Chart Using Discontinuous Columns

Let's make a bar chart of total government expenditures. The years are in Column A and the total government expenditures are in Column C, so they are in discontinuous columns. In this case, we can make a graph as follows: Select the data range of years (A1:A6) by dragging and move the mouse pointer to C1. (Do not click the mouse yet.) Holding the [Ctrl] key, drag C1 to C6. Then, we can select A1:A6 and C1:C6 (Fig. 3.19). We can use the same methods for columns separated by two or more columns and rows. Let's make the title **Fig. 4 Government Total Expenditures** and complete the graph.

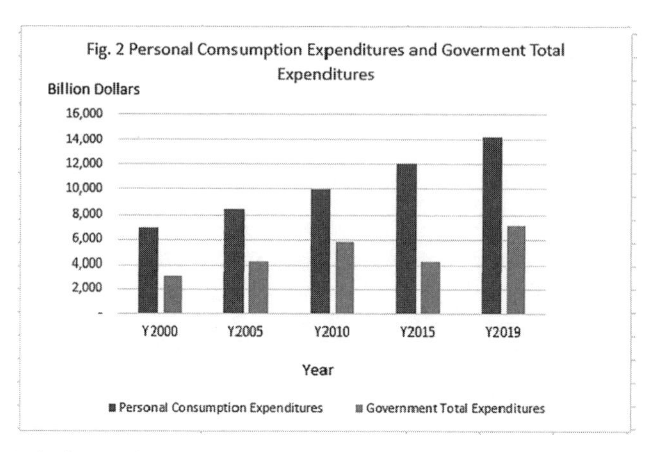

Fig. 3.16 Title the graph **Fig. 2 Personal Consumption Expenditures and Government Total Expenditures**, insert horizontal and vertical axis labels, and complete the graph.

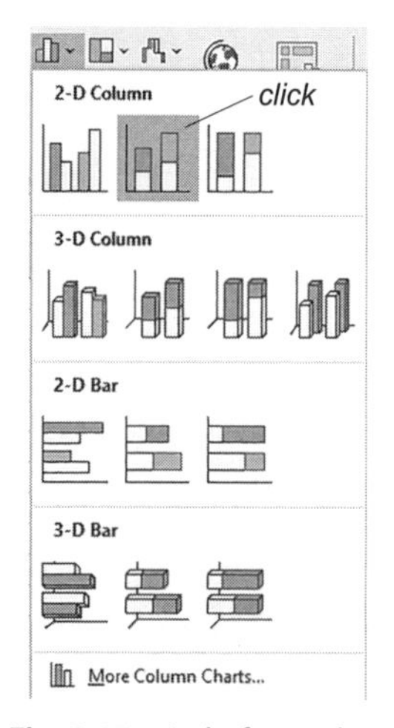

Fig. 3.17 As before, select the range by dragging. Click [Insert] → [Insert Column or Bar Chart] → [Stacked Column] under 2-D Column.

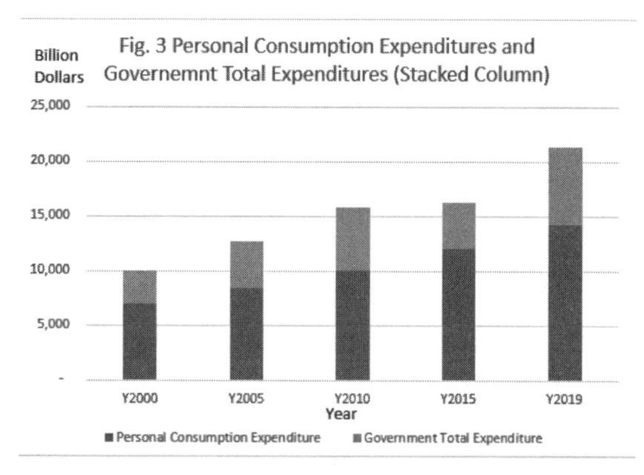

Fig. 3.18 Title the graph **Fig. 3 Personal Consumption Expenditures and Government Total Expenditures (Stacked Column)**, and insert the horizontal and vertical axis labels.

	A	B	C
1	Year	Personal Consumption Expenditures	Government Total Expenditures
2	Y2000	6,977	3,076
3	Y2005	8,470	4,196
4	Y2010	10,002	5,820
5	Y2015	12,040	4,196
6	Y2019	14,228	7,150

Fig. 3.19 Select the data range of years (A1:A6) by dragging and move the mouse pointer to C1. (Don't click the mouse yet.) Holding the [Ctrl] key, drag C1 to C6.

3.2 Inserting a New Worksheet

We have already made four graphs, and the current worksheet, Sheet1, is not easy to read. So, let's insert a new worksheet. In the [Home] tab, select the down arrow (∨) at the bottom of [Insert] in the [Cells] group, and choose [Insert Sheet] in the submenu. Then the new worksheet [Sheet2] is inserted (Fig. 3.20). Note that inserting a new sheet is done in the [Home] tab, not in the [Insert] tab. The active worksheet changes to [Sheet2]. When we want to come back to [Sheet1], click [Sheet1] at the lower right of the screen (Fig. 3.21). We can insert three or more sheets ([Sheet3], [Sheet4],...) by repeating the procedure. Let's copy the data in the range A1:C6 in [Sheet1] to the range starting from A1 in [Sheet2]. We can use [Copy/Paste] if the worksheets are different.

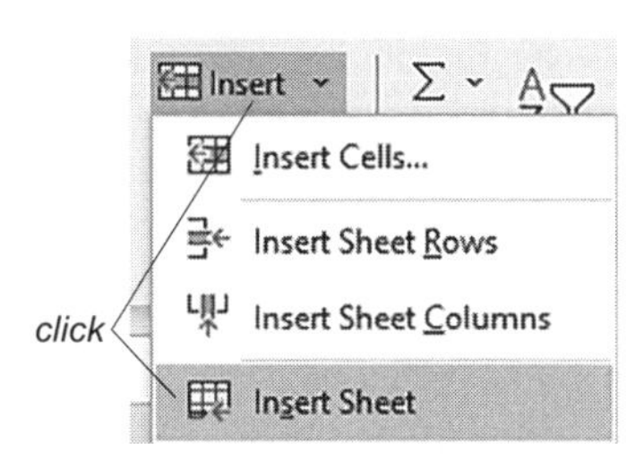

Fig. 3.20 On the [Home] tab, select the down arrow (∨) at the bottom of [Insert] in the [Cells] group, and choose [Insert Sheet] in the submenu.

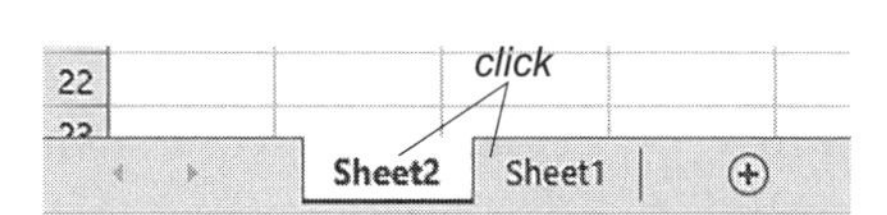

Fig. 3.21 When we want to come back to [Sheet1], click [Sheet1] at the lower right of the screen.

3.3 Pie Chart and Line Graph

3.3.1 Making a Pie Chart

Let's make a pie chart in [Sheet2]. Select the range B1:C2. Select [Insert] tab → [Insert Pie or Doughnut Chart] in the [Charts] group → [Pie] in [2-D Pie]. Change the title to **Fig. 5 Ratios of Personal Consumption and Government Total Expenditures**, move the graph to a proper position, and edit it to be able to read it easily.

Next, let's pull out a portion of the government total expenditures. Click the pie chart and make it active. Move the mouse pointer inside the portion of the government total expenditures that you want to pull out, and click once. Three round dots appear at the center and corners. Click the mouse once more, and drag the

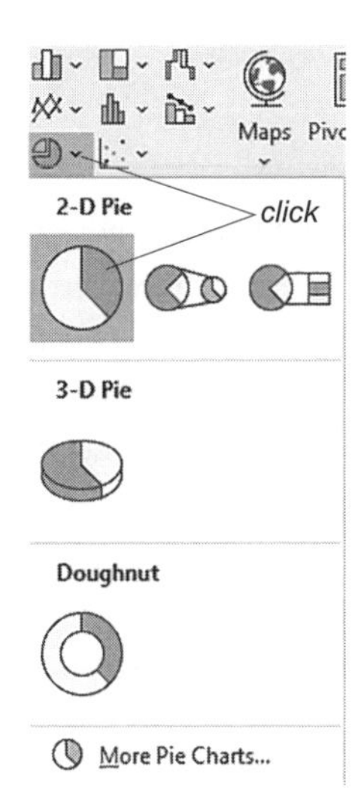

Fig. 3.22 Select [Insert] tab → [Insert Pie or Doughnut Chart] in the [Charts] group → [Pie] in [2-D Pie].

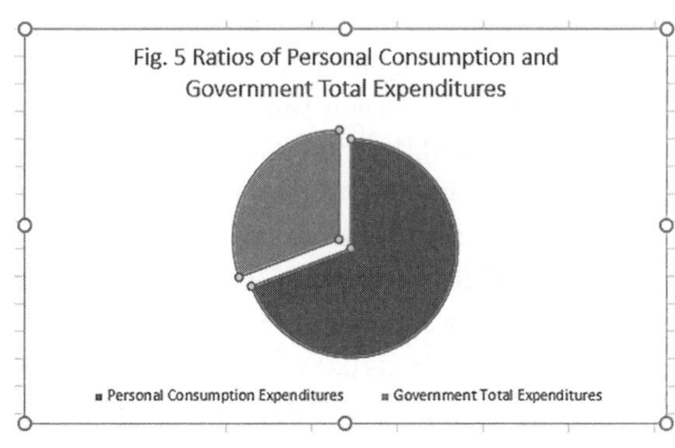

Fig. 3.23 Click the pie chart and make it active. Move the mouse pointer to inside the portion of Government Total Expenditures that you want to pull out, and click once. Three round dots appear at the center and corners. Click the mouse button once more, and drag the mouse pointer to outside of the circle. Then, the portion is pulled out.

mouse pointer outside of the circle. Then, the portion is pulled out (Fig. 3.23). If we want to return the portion back to its original position, drag it inside of the circle.

3.3.2 Line Graph

Let's make a line graph. Select the data ranges for the year and personal consumption expenditures, A1:B6. Click [Insert] → [Insert Line or Area Chart] → [Line] in [2-D Line] (Fig. 3.24). A line graph appears. Make the title **Fig. 6 Personal Consumption Expenditures (Line Graph)**. Edit the graph (inserting axis labels and changing the sizes) and move it to a proper place on the sheet to look good. If we need a graph involving multiple variables, select the range of the data, A1:C6, as in the bar chart case.

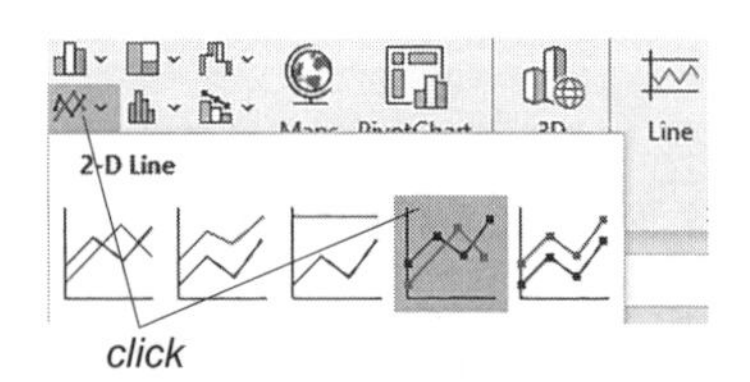

Fig. 3.24 Select the data range and click [Insert] → [Insert Line or Area Chart] → [Line] in [2-D Line].

3.4 Scatter Diagram (X-Y Graph)

In a line graph, data points appear equidistantly. However, the years are not equidistant. In cases where the datapoints are not equidistant or not ordered, we use a scatter diagram or an X-Y graph. For the scatter diagram, the horizontal axis must consist of numbers. Copy A1:C6 to the range starting from A11. Change the year data Y2000, Y2005,…, Y2019 (A12:A16) to numbers, namely 2000, 2005,…, 2019 (Fig. 3.25). Select the range of years and personal consumption expenditures, A11:B16. Click [Insert]→[Insert Scatter (X, Y) or Bubble Chart]→[Scatter with Straight Lines and Markers] (Fig. 3.26). The scatter diagram appears. Input the title, edit and move the graph, and complete it. If we want to use multiple variables, select the data range of variables as A11:C16. Note that the first column from the left is assigned to the horizontal (X) axis. So, if we want to change the horizontal (X) axis, we need to change the order of columns in the sheet.

Save the file as **EX3**, and finish the work.

	A	B	C
11	Year	Personal Consumption Expenditures	Government Total Expenditures
12	2000	6,977	3,076
13	2005	8,470	4,196
14	2010	10,002	5,820
15	2015	12,040	4,196
16	2019	14,228	7,150

Fig. 3.25 For the scatter diagram, the horizontal (X) axis must consist of numbers.

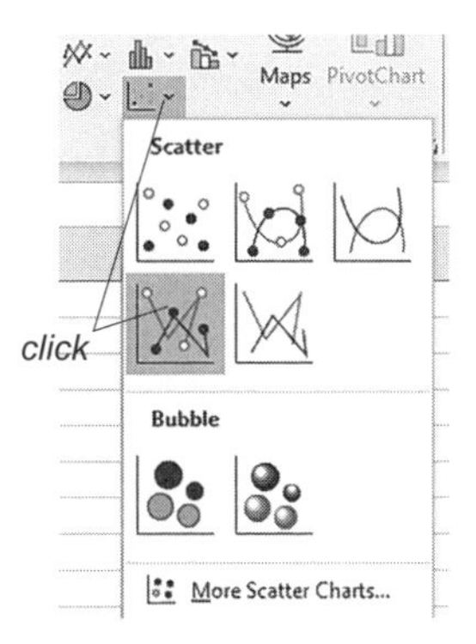

Fig. 3.26 Click [Insert]→[Insert Scatter (X, Y) or Bubble Chart]→[Scatter with Straight Lines and Markers].

3.5 Exercises Using World Population Data

Let's make various graphs using the world population data (Table 1.1 made in Chapter 1). Open the file, POP1, containing the world population data. Make following graphs.

1. A bar chart of the world population.
2. A clustered column chart of the populations of Asia and Africa.
3. A stacked column bar chart of all regions.
4. Pie charts of all regions in 1960 and 2017.
5. Pull out Africa from the pie charts made in 4.
6. A line graph of percentages of populations in Asia and Africa.
7. A scatter diagram of annual population growth rates in Asia and Africa.

4

Inputting a Large Dataset

4.1 Inputting Population Data of Countries

Table 4.1 shows the populations in 2020 and 2070 (unit: 1,000), areas (unit: 1,000 km^2), and GDP per capita ($US) in 2017 of major countries (the value for Venezuela is in 2014). The populations and areas were obtained from the medium estimates of "World Population Prospects: The 2017 Revision" and "Demographic Yearbook-2017" published by the United Nations. The GDP per capita data were obtained from "World Developing Indicators: 2019/4/24" published by the World Bank.

Hereafter, we do practices and exercises using this dataset. Let's input the data into Excel. A large dataset like this table cannot be displayed in its entirety on one screen. If we input the dataset using the methods we have studied, it is difficult to find the correct cells and easy to make mistakes. In this case, we use the Excel features of "Fill" and "Freeze Panes." Input the variable names **No.**, **Country**, **Region**, **GDP per capita**, **Area**, **Population 2020**, and **Population 2070** from A1 to G1, respectively. Do not input the title of the table to make the exercises easier. Since the necessary widths of the columns vary depending on the variables, change the sizes of the columns so that we can see the table easily. Since the numbers are input in the columns [No.], [GDP per capita], [Area], [Population 2020], and [Population 2070], change the alignment to right alignment in these columns (click [Right Alignment] in the [Home] tab).

Next, we input the data from the 2nd to 79th rows. The number of countries is 78, so we have to input numbers from 1 to 78. Excel has a feature for filling equally spaced numbers. Move the active cell to A2 and input the first number, **1**. When you press the [Enter] key, the

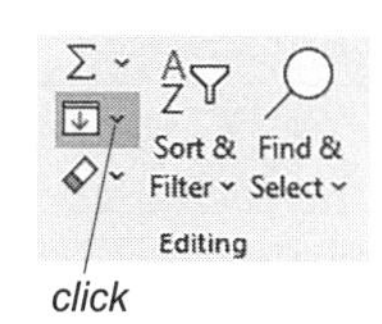

Fig. 4.1 Click [Fill] in the [Editing] group on the [Home] tab.

Table 4.1 Countries' Population Data.

No.	Country	Region	GDP per capita	Area	Population 2020	Population 2070
1	Algeria	Africa	4,055	2,382	43,333	61,690
2	Argentine	Latin America	14,398	2,796	45,510	58,221
3	Australia	Oceania	53,794	7,692	25,398	37,357
4	Austria	Europe	47,381	84	8,782	8,536
5	Bangladesh	Asia	1,517	148	169,775	199,365
6	Belgium	Europe	43,467	31	11,620	12,694
7	Bahamas	Latin America	3,394	1,099	11,544	17,574
8	Brazil	Latin America	9,812	8,516	213,863	221,895
9	Bulgaria	Europe	8,228	110	6,941	4,567
10	Burkina Faso	Africa	642	273	20,903	60,328
11	Cameroon	Africa	1,452	476	25,958	68,104
12	Canada	North America	44,871	9,985	37,603	48,240
13	Chile	Latin America	15,346	756	18,473	20,414
14	China	Asia	8,827	9,600	1,424,548	1,208,909
15	Colombia	Latin America	6,409	1,142	50,220	52,049
16	Congo, Democratic Republic of	Africa	463	2,345	89,505	280,414
17	Cote d'Ivoire	Africa	1,538	322	26,172	72,573
18	Denmark	Europe	57,219	43	5,797	6,569
19	Ecuador	Latin America	6,273	257	17,336	24,670
20	Egypt	Africa	2,413	1,002	102,941	178,407
21	Ethiopia	Africa	768	1,104	112,759	229,097
22	Finland	Europe	45,805	337	5,580	6,019
23	France	Europe	38,484	552	65,721	71,956
24	Germany	Europe	44,666	358	82,540	75,164
25	Ghana	Africa	2,046	239	30,734	63,943
26	Greece	Europe	18,885	132	11,103	8,670
27	Guatemala	Latin America	4,471	109	17,911	30,867
28	Hungary	Europe	14,279	93	9,621	7,365
29	India	Asia	1,979	3,287	1,383,198	1,665,179
30	Indonesia	Asia	3,846	1,911	272,223	323,653
31	Iran	Asia	5,594	1,629	83,587	87,177
32	Ireland	Europe	68,885	70	4,888	6,052
33	Israel	Asia	40,544	22	8,714	14,803
34	Italy	Europe	32,110	302	59,132	50,533
35	Japan	Asia	38,430	378	126,496	96,369
36	Kazakhstan	Asia	9,030	2,725	18,777	24,595
37	Kenya	Africa	1,595	592	53,492	120,634
38	Korea, Republic of	Asia	29,743	100	51,507	44,925
39	Kuwait	Asia	29,040	18	4,303	5,967

40	Madagascar	Africa	450	587	27,691	73,274
41	Malawi	Africa	338	118	20,284	57,643
42	Malaysia	Asia	9,952	330	32,869	43,698
43	Mali	Africa	827	1,240	20,284	62,163
44	Mexico	Latin America	8,910	1,964	133,870	166,496
45	Morocco	Africa	3,023	447	37,071	46,843
46	Mozambique	Africa	426	799	32,309	96,544
47	Nepal	Asia	849	147	30,260	35,591
48	Netherlands	Europe	48,483	42	17,181	17,075
49	New Zealand	Oceania	42,583	268	4,834	5,991
50	Niger	Africa	378	1,267	24,075	115,399
51	Nigeria	Africa	1,968	924	206,153	576,062
52	Norway	Europe	75,704	324	5,450	7,461
53	Pakistan	Asia	1,548	796	208,362	343,516
54	Peru	Latin America	6,572	1,285	33,312	43,124
55	Philippines	Asia	2,989	300	109,703	167,443
56	Poland	Europe	13,864	313	37,942	27,360
57	Portugal	Europe	21,291	92	10,218	7,789
58	Romania	Europe	10,819	238	19,388	14,260
59	Russia	Europe	10,749	17,098	143,787	126,393
60	Saudi Arabia	Asia	20,849	2,207	34,710	46,160
61	Singapore	Asia	57,714	1	5,935	6,215
62	South Africa	Africa	6,151	1,221	67,595	91,959
63	Spain	Europe	28,208	506	46,459	39,843
64	Sri Lanka	Asia	5,317	66	21,084	18,693
65	Sweden	Europe	53,253	439	10,122	12,444
66	Switzerland	Europe	80,343	41	8,671	10,195
67	Tanzania	Africa	958	947	62,775	204,040
68	Thailand	Asia	6,595	513	69,411	57,438
69	Tunisia	Africa	3,464	164	11,903	13,981
70	Turkey	Asia	10,546	780	83,836	94,970
71	Ukraine	Europe	2,640	604	43,579	31,992
72	United Kingdom	Europe	39,954	242	67,334	78,212
73	United States of America	North America	59,928	9,834	331,432	419,162
74	Venezuela	Latin America	15,692	930	33,172	43,357
75	Viet Nam	Asia	2,342	331	98,360	114,496
76	Yemen	Asia	1,107	528	30,245	54,320
77	Zambia	Africa	1,513	753	18,679	61,286
78	Zimbabwe	Africa	1,333	391	17,680	36,164

Sources; Populations (1,000) and areas (1,000 km^2): the medium estimates of "World Population Prospects: The 2017 Revision" and "Demographic Yearbook-2017" published by the United Nations; GDP per capita data ($US): "World Developing Indicators: 2019/4/24" published by the World Bank.

Fig. 4.2 Choose [Series] in the submenu.

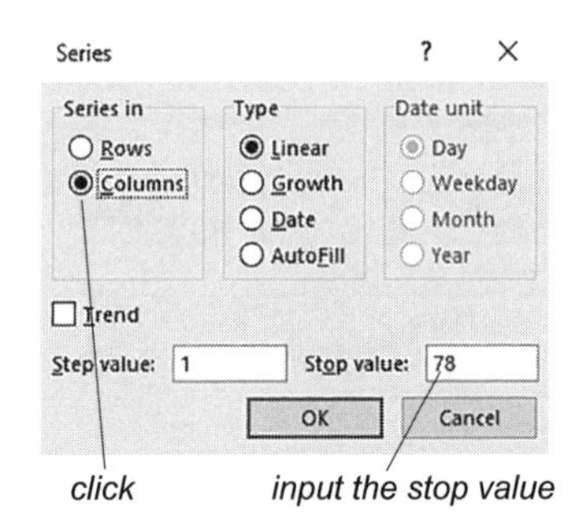

Fig. 4.3 Set [Series in] to [Columns] and set the [Stop value] as **78**.

active cell moves down to A3. Move the active cell to A2. Click [Fill] in the [Editing] group on the [Home] tab (Fig. 4.1). Choose [Series] in the submenu (Fig. 4.2), and the [Series] box appears. Choose [Columns] under [Series in],and make the [Stop value] **78** (Fig 4.3). When we click [OK], the numbers 1 to 78 are automatically input from A2 to A79. Input the names of the countries from B2 to B79.

Next, we input other variables. But if you input Table 4.1 as it is, it is difficult to input correctly. So, we fixed the panes so that the row of variable names (Row 1) and the [No.] and [Country] columns (Columns A and B) always appear independent of the position of the active cell. Move the active cell C2. Click the [View] tab and select [Freeze Panes] in the [Window] group. Choose [Freeze Panes] in the submenu (Fig 4.4). The panes are fixed, and the variable names, [No.], and [Country] are displaced. If you want to stop the panes from being fixed, select [Freeze Panes] → [Unfreeze Panes] in the [View] tab (Fig. 4.5). Let's input all of the data. Be careful to input the data for [Region] in the same style. (For example, use the same lower and upper case styles and the same space for-

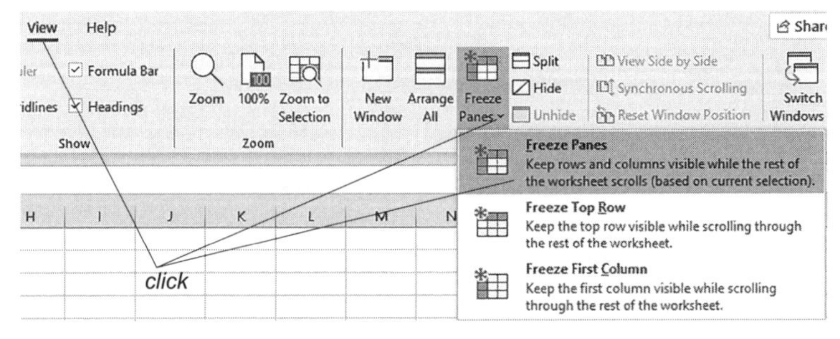

Fig. 4.4 Move the active cell to C2. Click the [View] tab and select [Freeze Panes] in the [Window] group.

mat. If you input one space in some cells (i.e., "Latin ␣ America") and two or more spaces (i.e.,"Latin ␣ ␣ America") in other cells (a space is represented by the ␣ symbol), we cannot do the practices and exercises correctly. Stop the panes from being fixed, change the format of the numbers to the comma style, and use ruled lines to be able to read the table easily.

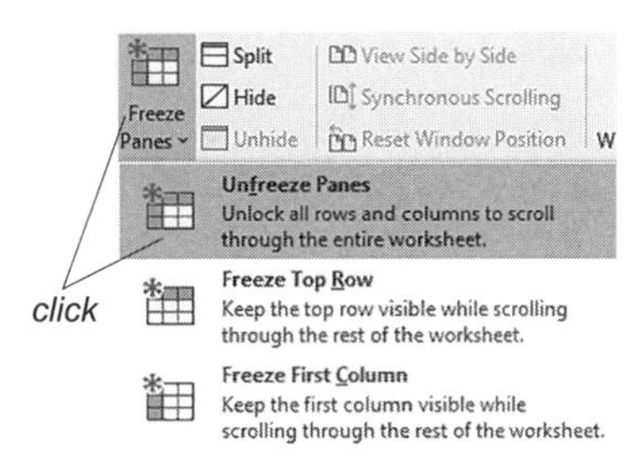

Fig. 4.5 To set the panes so that they are not fixed, select [Freeze Panes] → [Unfreeze Panes] in the [View] tab.

The data we have input are classified into two categories. One category includes variables that take numerical values, such as population, area, and GDP per capita. Such data are called "quantitative data." The other category includes variables such as the country names and regions. Such data are called "qualitative data." In statistics, various methods have been developed for analyzing not only quantitative but also qualitative data. Save the file as **POP2**. Hereafter, we refer to these data as "Countries' Population Data."

4.2 Exercises Calculating Population Densities and Population Growth Rates

Let's calculate the population densities for 2020 in Column H and the Population Growth Rates in Column I.

Input **Population Density** in H1 and **Population Growth Rate** in I1. Then calculate the population densities from H2 and the population growth rates from I2. Check Chapter 2 for the actual calculation methods.

Change the column widths, change the format of the population densities to the comma style with one digit after the decimal point, and change the format of the population growth rates to the percent style with two digits after the decimal point. Draw ruled lines and complete the table.

5

Sorting the Data and Obtaining the Data Matching Criteria

In this chapter, we sort the data and obtain the data that match given criteria. Open the file containing the Countries' Population Data that we input in the previous chapter. The names of the variables [No.], [Countries],…, [Population Growth Rate] in the first row, are referred to as "field names." The data are input from the 2nd to the 79th row, and the range containing both the field names and data are called the "List". We perform various operations such as sorting and selecting data using the List. To perform these operations correctly, it is important to note the following:

i) The data start just below the field names; there is no blank row between the field names and the data.
ii) In the data set, there are no blank rows or columns.
iii) The List is separated by one or more rows and columns from other Lists and from other cells not belonging to the List.

5.1 Sorting the Data

Let's sort the population data in descending order (largest to smallest) by Population 2020. Let's specify the range of the List (field names + data) using the mouse. We can specify the whole range of the List (A1:I79); however, it is difficult to specify the whole range when the size of the List is large, as in this example. So, in Excel, we can specify the List if we move the active cell to one (just one) cell in the List. Move the active cell to any cell in the List.

Next, click the [Data] tab and choose [Sort] (Fig. 5.1). The Sort box appears. Click the downward arrow for the [Sort by] box under [Column]. Field names are displayed. Choose [Population 2020] (Fig. 5.2). Then, click the downward arrow for the [Smallest to Largest] box under [Order]. Then, select [Largest to Smallest] (Fig. 5.3, default is [Smallest to Largest]). Click [OK], and the data set is sorted

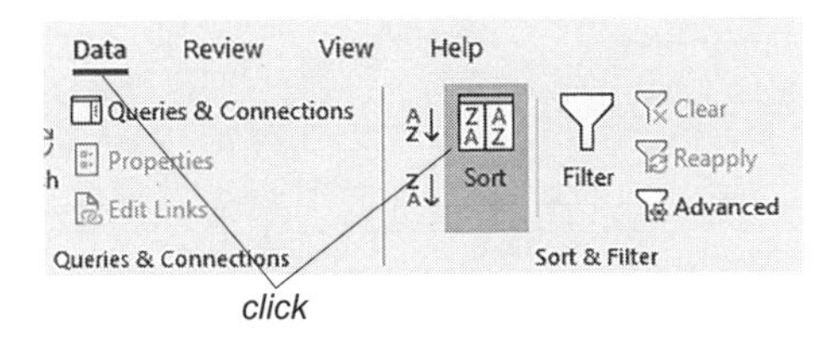

Fig. 5.1 Click the [Data] tab, and choose [Sort]

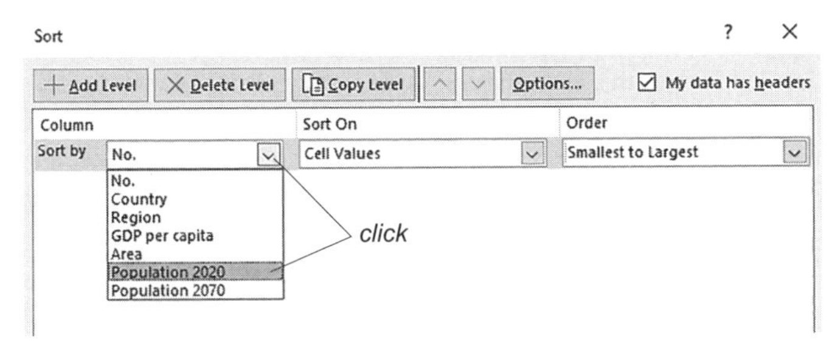

Fig. 5.2 Click the downward arrow of the [Sort by] box under [Column]. Field Names are displayed. Choose [Population 2020].

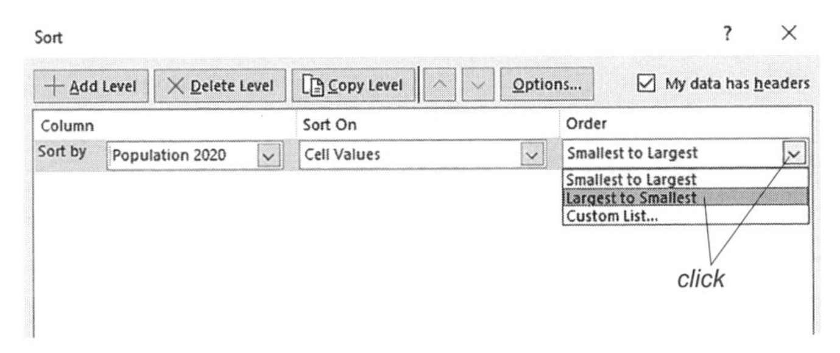

Fig. 5.3 Click the downward arrow of the [Smallest to Largest] box under [Order]. Then, select [Largest to Smallest] (default is [Smallest to Largest]). Click [OK]. The data set is sorted in descending order by [Population 2020].

in descending order by [Population 2020]. We can also sort character data by alphabetic order. In the [Sort] box, if we click [Add Level], [Then by] is added and we can do a detailed sort. In this case, the sort is done in lexicographic order; this means that the sort is first done by [Sort by] and then [Then by] if the same values are present multiple times. We can add more [Then by] criteria by clicking [Add Level]. When we want to delete the level, click [Delete Level] (Fig. 5.4).

In our List, the field names and the data are different types of entries (field names are treated as characters and the data are numbers and characters.) However, when the field name and data are the same type of entry, for example, when

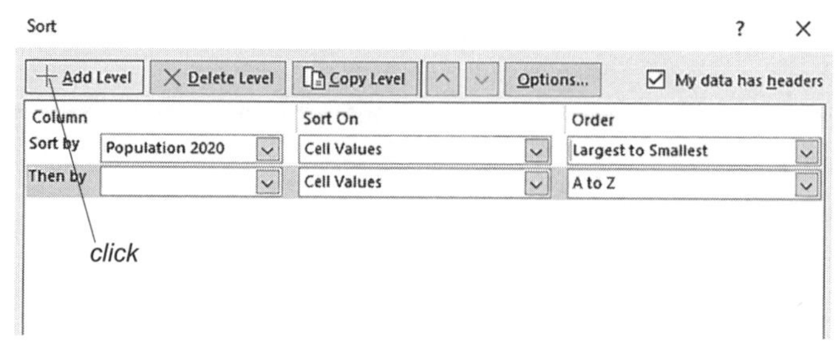

Fig. 5.4 If we click [Add Level], [Then by] is added, and we can do a detailed sort. In this case, the sort is done by lexicographic order. This means that the sort is first done by [Sort by] and then [Then by] if the same values are present. We can add more [Then by] criteria by clicking [Add Level]. When we want to delete a level, click [Delete Level].

we sort the names of our friends, we have to be careful. Excel may include the field names as part of the data when it sorts. In this case,

i) Click the box for [My data has headers] so that the box is checked.
ii) Specify the range of the data only (not including field names) and sort the data (in this case you need to specify the entire data range such as A2:I79, not just one cell).

Let's sort by [No.] and return the List to the original version.

5.2 Obtaining Data Matching the Criteria

5.2.1 Using [Filter]

Let's obtain the data that match given criteria. First, obtain the countries whose Population 2020 values are greater than or equal to 50 million. Move the active cell to a cell in the List (field names + data). Next, choose the [Data] tab, and click [Filter] in the [Sort & Filter] group (Fig. 5.5). Then, downward arrows appear on the right sides of the

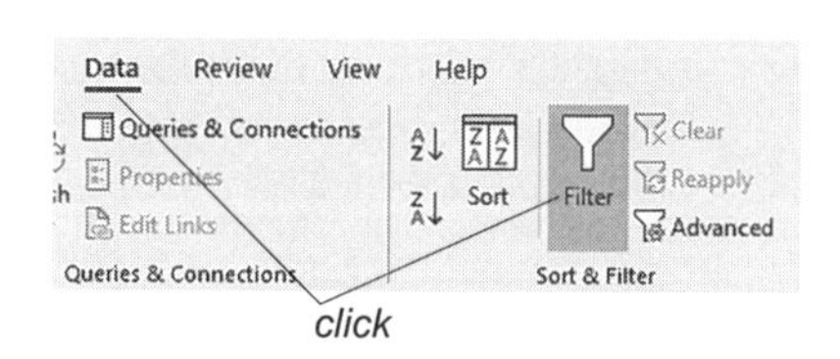

Fig. 5.5 Choose the [Data] tab and click [Filter] in the [Sort & Filter] group.

	A	B	C	D	E	F	G
1	N	Country	Region	GDP per cap	Ar	Population 20	Population 20
2	1	Algeria	Africa	4,055	2,382	43,333	61,690
3	2	Argentine	Latin America	14,398	2,796	45,510	58,221
4	3	Australia	Oceania	53,794	7,692	25,398	37,357
5	4	Austria	Europe	47,381	84	8,782	8,536

Fig. 5.6 Downward arrows appear on the right sides of the Field Names.

field names (Fig. 5.6). Click the downward arrow for [Population 2020]; the menu of the Filter appears. Choose [Number Filters] → [Greater Than Or Equal To] (Fig. 5.7). The box [Custom AutoFilter] appears. Since population data are given in units of one thousands, we input **50000** for 50 million and click [OK] (Fig. 5.8). The countries whose Population 2020 values are greater than or equal to 50 million are displayed, and the cell numbers change to blue. (But all data are stored in the PC). Next, add more criteria to obtain the countries whose Population 2020 values are over 50 million and in Asia and Africa. Click the downward arrow for [Region] → [Text Filters]→[Equals…] (Fig 5.9). The box [Custom

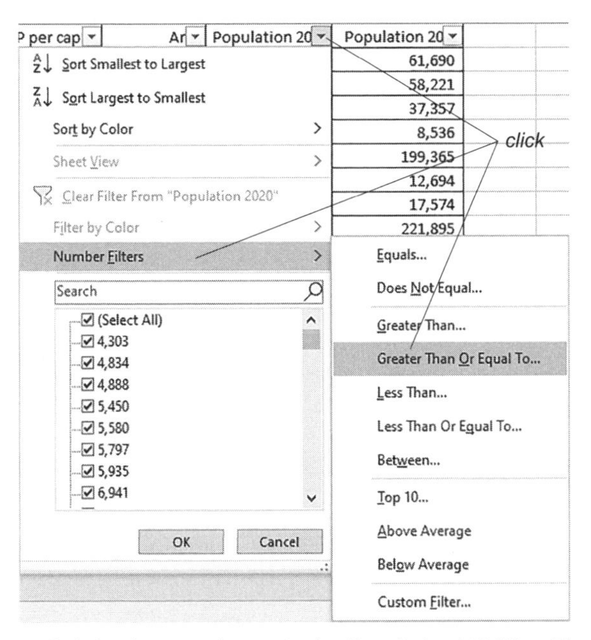

Fig. 5.7 Click the downward arrow in the [Population 2020] box. The Filter menu appears. Choose [Number Filters] → [Greater Than Or Equal To].

Custom AutoFilter ? ×

Show rows where:
Population 2020
is greater than or equal to | 50000
And Or
Use ? to represent any single character
Use * to represent any series of characters
OK Cancel

Fig. 5.8 Input **50000** for 50 million and click [OK].

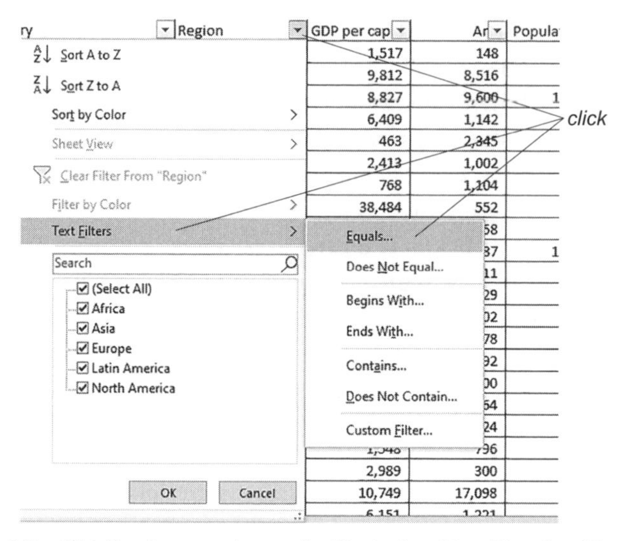

Fig. 5.9 Click the downward arrow for [Region] → [Text Filters] → [Equals].

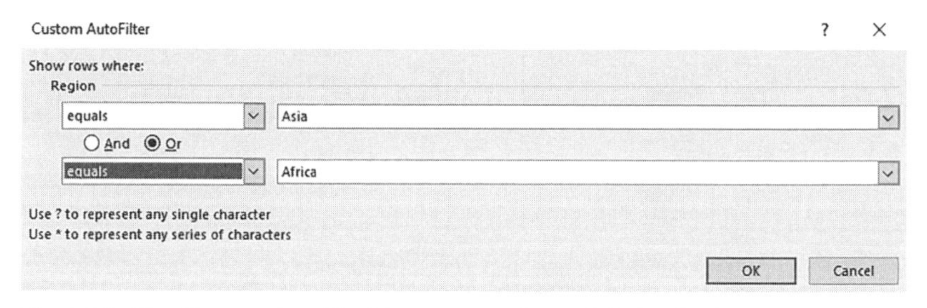

Fig. 5.10 The [Custom AutoFilter] box appears. Input **Asia** for the first criterion. Since the countries may be in either Asia or Africa, click [Or]. Choose [equals] and input **Africa** as the second criterion. Click [OK].

	A	B	C	D	E	F	G	H	I
1	N	Country	Region	GDP per cap	Ar	Population 20	Population 20	Population Densi	Population Growth Ra
6	5	Bangladesh	Asia	1,517	148	169,775	199,365	1,147.1	0.32%
15	14	China	Asia	8,827	9,600	1,424,548	1,208,909	148.4	-0.33%
17	16	Congo, Democratic Republic of	Africa	463	2,345	89,505	280,414	38.2	2.31%
21	20	Egypt	Africa	2,413	1,002	102,941	178,407	102.7	1.11%
22	21	Ethiopia	Africa	768	1,104	112,759	229,097	102.1	1.43%
30	29	India	Asia	1,979	3,287	1,383,198	1,665,179	420.8	0.37%
31	30	Indonesia	Asia	3,846	1,911	272,223	323,653	142.5	0.35%
32	31	Iran	Asia	5,594	1,629	83,587	87,177	51.3	0.08%
36	35	Japan	Asia	38,430	378	126,496	96,369	334.6	-0.54%

Fig. 5.11 The countries with a [Population 2020] value over 50 million in Asia or Africa are displayed.

AutoFilter] appears; input **Asia** for the first criterion. Since the countries may be either Asia or Africa, click [Or]. Choose [equals] and input **Africa** as the second criterion. Click [OK] (Fig. 5.10). Then, the countries with a Population 2020 value over 50 million in Asia or Africa are displayed (Fig. 5.11).

If you are finished using "Filter" click [Filter] in the [Data] tab (Fig. 5.12).

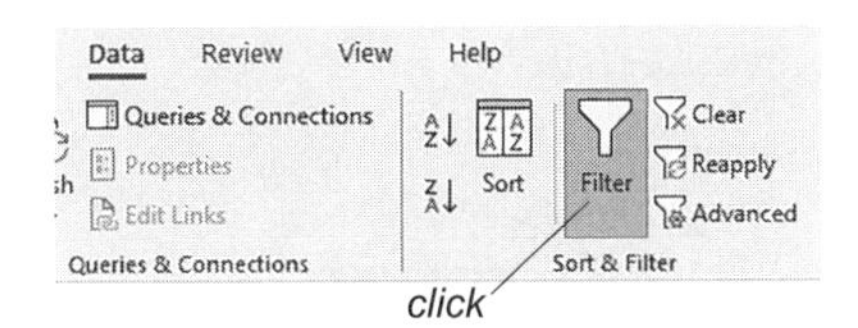

Fig. 5.12 If you are finished setting the "Filter" click [Filter] in the [Data] tab.

5.2.2 Using [Advanced] in [Sort & Filter]

If the criteria are simple, we can obtain the data that match the criteria using [Filter]. However, if the criteria are complicated, we use [Advanced] in [Sort & Filter]. Make sure [Filter] is off (if not, click it to close it), and the [Data] tab is selected.

a. Obtaining All Variables

We want to obtain the countries with a [Population 2020] value over 50 million. To obtain the data using [Advanced] in [Sort & Filter], we need to input the criterion in the sheet. Let's input the criterion from K2. Copy the field name [Population 2020] in K2 from the List. The field name must be the exactly same as one in the List, so copy it. In K3, just below K2, input the criterion **>50000** (Fig. 5.13). The criteria and the symbols representing them include:

= Equals, > Greater Than, >= Greater Than Or Equal To, < Less Than, <= Less Than Or Equal To, and <> Does Not Equal.

After setting the criterion, we obtain the countries with a Population 2020 value over 50 million. Move the active cell to a cell in the List of the population data. Choose [Advanced] in [Sort & Filter] on the [Data] tab (Fig. 5.14). The [Advanced Filter] box appears; check that [List range] is correctly ($A$1:$I$79) selected. Next, we set the [Criteria range]. We can type in the range (**K2:K3**), but we can also set it using the mouse. Click the upward arrow to the right of [Criteria range] (Fig. 5.15). The [Advanced Filter -Crit…] box appears. Set the [Criteria range] to K2:K3 by dragging with the mouse. After setting the [Criteria range], click the

	J	K	L
1			
2		Population 2020	
3		>50000	
4			
5			

Fig. 5.13 Input the criteria in K2:K3.

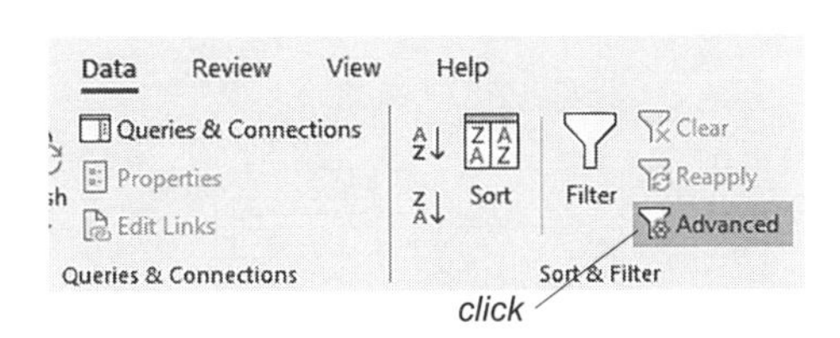

Fig. 5.14 Choose [Advanced] in the [Sort & Filter] area of the [Data] tab.

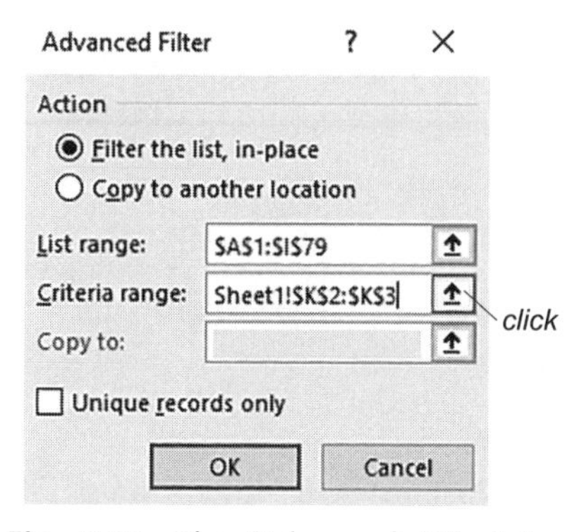

Fig. 5.15 The [Advanced Filter] box appears. Check that the [List range] is correctly ($A$1:$I$79) selected. We set the [Criteria range]. We can type it (**K2:K3**) or we can set it using the mouse. Click the upward arrow to the right of [Criteria range].

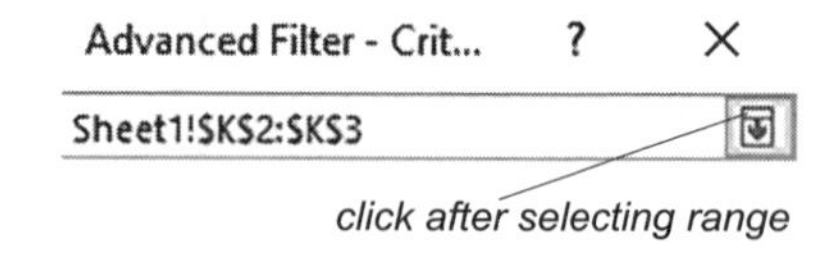

Fig. 5.16 The [Advanced Filter -Crit…] box appears. Set the [Criteria range] to K2:K3 by dragging. After setting the [Criteria range], click the down arrow to the right of the box.

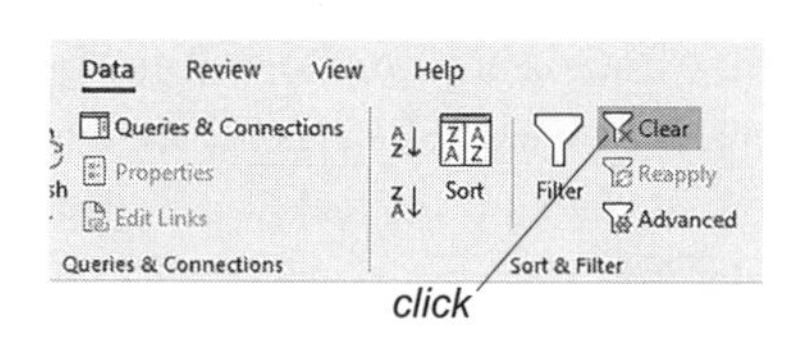

Fig. 5.17 To finish displaying the results, click [Clear] in [Sort & Filter].

down arrow to the right of the box (Fig. 5.16). We go back to the [Advanced Filter] box and click [OK]. A matching result appears on the display as the [Filter] case. When finished displaying the results, click [Clear] in [Sort & Filter] (Fig. 5.17).

Next, write the matching results on the worksheet. Let's write the results from K6. As before, move the active cell to a cell in the List of population data. Choose [Advanced] in [Sort & Filter]. The [Advanced Filter -Crit…] box appears. Check that the ranges of the List and Criteria are correct (if they are not, correct them). Choose [Copy to another location] and input

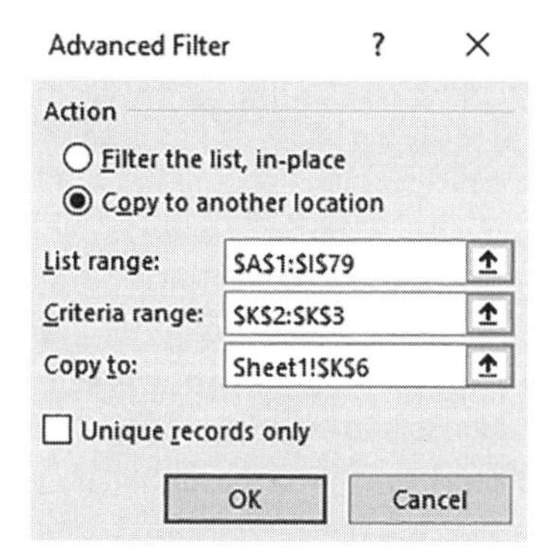

Fig. 5.18 Choose [Copy to another location] and input **K6** in the [Copy to:] box by typing or using the mouse.

	K	L	M	N	O	P	Q	R	S
6	No.	Country	Region	GDP per capita	Area	Population 2020	Population 2070	Population Density	Population Growth Rate
7	5	Bangladesh	Asia	1,517	148	169,775	199,365	1,147.1	0.32%
8	8	Brazil	Latin America	9,812	8,516	213,863	221,895	25.1	0.07%
9	14	China	Asia	8,827	9,600	1,424,548	1,208,909	148.4	-0.33%
10	15	Colombia	Latin America	6,409	1,142	50,220	52,049	44.0	0.07%
11	16	Congo, Democratic Republic of	Africa	463	2,345	89,505	280,414	38.2	2.31%
12	20	Egypt	Africa	2,413	1,002	102,941	178,407	102.7	1.11%

Fig. 5.19 The matching result is output from K6.

K6 in the box [Copy to:] by typing or using the mouse (Fig. 5.18). Click [OK], and the matching result is then output from K6 (Fig 5.19). Note that we only input one cell in [Copy to:]

b. Obtain Selected Fields

Although we have obtained all of the fields in the previous section, we often do not need all of the fields. Here, we obtain the data for just three fields: [Country], [Region], and [Population 2020]. Delete the results obtained in the previous section from K6 by dragging the range and clicking [Clear] in the [Editing] group of the [Home] tab → [Clear All] (Fig. 5.20). Next, copy the field names [Country], [Region], and [Population 2020] to the cells from K6 to M6. Move the active cell in the List, and select the [Data] tab → [Advanced] in [Sort & Filter]. Make sure that the [List range] and [Criteria range] are correct. Click [Copy to another location], delete the previous range in [Copy to], and input **K6:M6** (Fig. 5.21). Click [OK] and the [Country], [Region], and [Population 2020] data are obtained from K6 (Fig. 5.22).

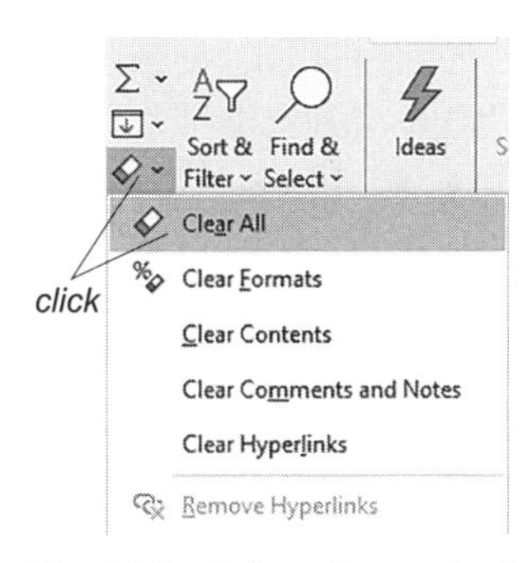

Fig. 5.20 Delete the result of the previous section from K6 by dragging the range and click [Clear] in the [Editing] group of the [Home] tab → [Clear All].

Advanced Filter ? ×

Action

○ Filter the list, in-place

◉ Copy to another location

List range: $A$1:$I$79

Criteria range: $K$2:$K$3

Copy to: Sheet1!$K$6:$M$6

☐ Unique records only

OK Cancel

Fig. 5.21 Move the active cell in the List, and select the [Data] tab → [Advanced] in [Sort & Filter]. Click [Copy to another location], delete the previous range in [Copy to], and input **K6:M6**.

	K	L	M
6	Country	Region	Population 2020
7	Bangladesh	Asia	169,775
8	Brazil	Latin America	213,863
9	China	Asia	1,424,548
10	Colombia	Latin America	50,220
11	Congo, Democratic Republic of	Africa	89,505

Fig. 5.22 The data for [Country], [Region], and [Population 2020] are obtained from K6.

c. Obtaining Data Matching Multiple Criteria

Here, let's obtain data matching multiple criteria. We obtain the countries in which the value of [Population 2020] is more than 50 million and less than or equal to 100 million and that are also in Asia or Africa. First, delete the previous criteria and the output results. Next, we input the matching criteria. Copy [Region] from the field names to K2, and copy [Population 2020] to L2 and M2. To select for countries with a Population 2020 value more than 50 million and less than or equal to 100 million, we need to input two criteria.

Input **Asia** in K3, **>50000** in L3, and **<=100000** in M3. Then, input **Africa** in K4, **>50000** in L4, and **<=100000** in M4. Do not input blank cells in the criteria range. Asia and Africa must be exactly the same as those in the List (Fig. 5.23).

In Excel, if the criteria are in the same row, the data that satisfy all of the criteria in the row are selected (using the "and" feature). If the criteria are in different rows, the data satisfying at least one row (using the "or" feature) are selected. The cells from K3 to M3 are in the same row, so the matching criteria are countries in Asia and countries with a 2020 Population value more than 50 million and less than or equal to 100 million. In the same way, the criteria from K4 to L4 are Africa and a Population 2020 value more than 50 million and less than or equal to 100 million. Since these criteria are in different rows, the matching data satisfy one of these rows and become countries in Asia or Africa with a Population 2020 value more than 50 million and less than or equal to 100 million.

Let's obtain the data matching the criteria. Select [Data] tab → [Advanced] in [Sort & Filter]. The [Advanced Filter] box appears. Check the [List range]. Then,

	K	L	M
1			
2	Region	Population 2020	Population 2020
3	Asia	>50000	<=100000
4	Africa	>50000	<=100000

Fig. 5.23 Input **Asia** in K3, **>50000** in L3, and **<=100000** in M3. Then, input **Africa** in K4, **>50000** in L4, and **<=100000** in M4. Do not input blank cells in the criteria range. The entries "Asia" and "Africa" must be be identical the entries for "Asia" and "Africa" in the List.

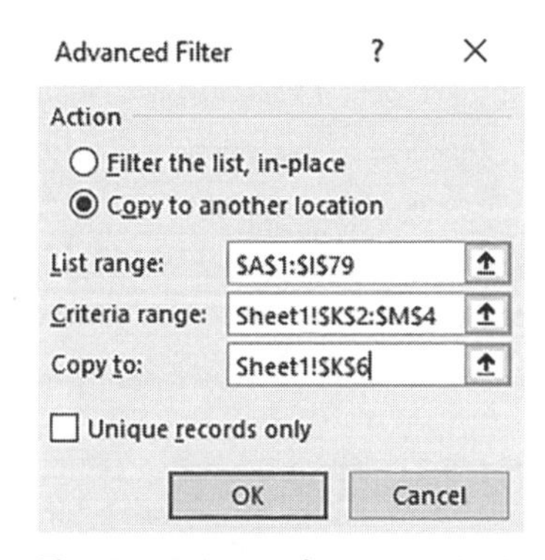

Fig. 5.24 Input the new criteria range (**K2:M4**) in [Criteria range:]. Click [Copy to another location:] → input **K6** in [Copy to:] → click [OK].

	K	L	M	N	O	P	Q	R	S
6	No.	Country	Region	GDP per capita	Area	Population 2020	Population 2070	Population Density	Population Growth Rate
7	16	Congo, Democratic Republic of	Africa	463	2,345	89,505	280,414	38.2	2.31%
8	31	Iran	Asia	5,594	1,629	83,587	87,177	51.3	0.08%
9	37	Kenya	Africa	1,595	592	53,492	120,634	90.4	1.64%
10	38	Korea, Republic of	Asia	29,743	100	51,507	44,925	515.1	-0.27%
11	62	South Africa	Africa	6,151	1,221	67,595	91,959	55.4	0.62%
12	67	Tanzania	Africa	958	947	62,775	204,040	66.3	2.39%
13	68	Thailand	Asia	6,595	513	69,411	57,438	135.3	-0.38%
14	70	Turkey	Asia	10,546	780	83,836	94,970	107.5	0.25%
15	75	Viet Nam	Asia	2,342	331	98,360	114,496	297.2	0.30%

Fig. 5.25 The data for countries in Asia or Africa with a Population 2020 value more than 50 million and less than or equal to 100 million are obtained.

delete the already input range (from K2 to K3) in [Criteria range:] and input the new criteria range (**K2:M4**). Next, click [Copy to another location:] → input **K6** in [Copy to:] → click [OK] (Fig. 5.24). The data for countries in Asia or Africa with a 2020 population more than 50 million and less than or equal to 100 million are obtained (Fig. 5.25).

When we want to obtain the data in some fields, then copy the field names and specify them as the output location as before.

5.3 Summarizing the Data Matching Criteria Using Database Functions

We can easily get summaries of the data matching criteria using database functions without obtaining the data that match the criteria. Let's summarize the data for Asian countries using the database functions. To avoid unnecessary complications, delete the criteria and results we have made. To consider Asian countries, copy the field name, [Region], to K2 and input **Asia** in K3.

Here, we calculate several different values that summarize the data. It is inconvenient to specify the ranges using the cell addresses. So, we can name the List and criteria ranges, and use the names to summarize the data. First, name the List. Select the ranges on the List (A1:I79) by dragging using the mouse. Click the [Formulas] tab → [Define Name] in the [Define Names] group (Fig. 5.26). The [New Name] box appears, so make [Name] **Data1** and click [OK] (Fig. 5.27). In the same way, name the region for the criteria (K2:K3) **Criteria1**. We can use these names to specify the ranges.

Next, let's calculate the sum of the Population 2020 values in Asian countries. Move the active cell to K6 and input **=DSUM(Data1, "Population 2020", Criteria1)**. The sum of the Population 2020 values in Asia is calculated. DSUM may be given either in upper or lower case letters. Instead of "Population 2020," we

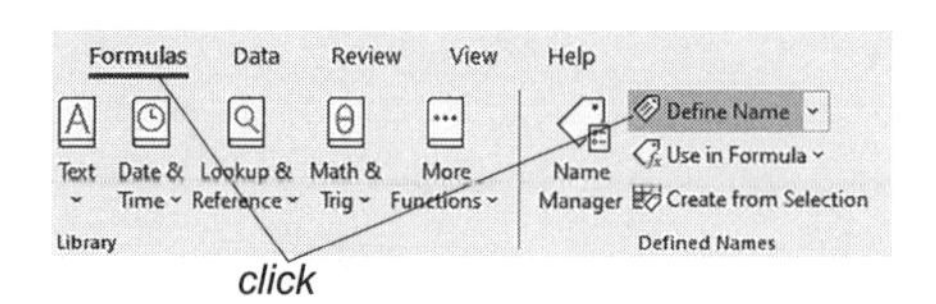

Fig. 5.26 Select the ranges of the List (A1:I79) by dragging using the mouse. Click the [Formulas] tab → [Define Name] in the [Define Names] group.

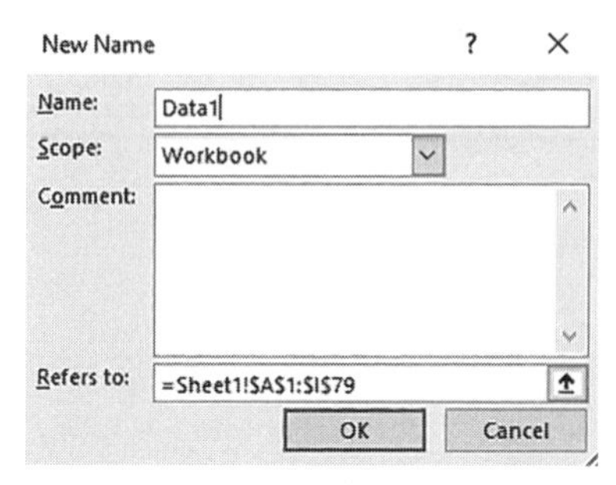

Fig. 5.27 The [New Name] box appears, so make the [Name] **Data1** and click [OK].

	K	L	M
1			
2	Region		
3	Asia		
4			
5			
6	=DSUM(Data1, "Population 2020", Criteria1)		
7			

Fig. 5.28 The sum of the [Population 2020] values of countries in Asia is calculated by the database function.

can also use the cell address and input **=DSUM(Data1, F1, Criteria1)**. The database functions begin with "D." When we use the database functions, three arguments are required, and they are:

i) Range of the List
ii) Summarized field name surrounded by double quotations " " (or its cell address)
iii) Range of the criteria

The three arguments must be in this order and must be separated by commas, and we write them as follows: Database Function Name (List Range, Field Name, Criteria Range).

The database functions are:

DAVERAGE average
DCOUNT number of cells that contain numbers
DCOUNTA number of nonempty cells
DGET A value that matches the criteria
DMAX Maximum
DMIN Minimum

DPRODUCT Product
DSTEDEV Standard deviation
DSUM Sum
DVAR Variance

Using these functions, let's calculate the average [GDP per capita] and maximum [Population Growth Rate] in Asia. Input **=DAVERAGE(Data1, "GDP per capita", Criteria1)** in K7 and **=DMAX(Data1, "Population Growth Rate", Criteria1)** in K8. Then we calculate the average of [GDP per capita] and the maximum of [Population Growth Rate] in Asia.

If the criteria are complicated, we set the criteria using the same methods previously studied in this chapter. Note that if the criteria are changed, the values of the database functions will change. In this case, you need to copy the values in the cells as explained in Chapter 2.

5.4 Exercises Sorting and Obtaining Matched Data Using the Population Data

Using the Countries' Population Data, try the following sorting tasks. Then obtain the data matching the criteria and calculate the vlaues using the data base functions. In the exercises, do not delete the results, and keep them in the proper ranges from the sheet.

1. Find the 10 largest population countries in 2020 and 2070, and compare them.
2. Find the 10 countries whose population growth rates are the largest.
3. Obtain [Country] and [Population Growth Rate] for the countries whose Population Growth Rates are greater than 1%.
4. Obtain [Country], [Region], [GDP per capita], and [Population Growth Rate] for the Asian or African countries whose population growth rates are greater than 0.5% and less than or equal to 1%.
5. Using the database function, calculate the average and minimum [Population Growth Rate] in countries whose GDP per capita is (i) less than $5,000 and (ii) greater than or equal to $7,000.
6. Using the database function, calculate the average and maximum [GDP per capita] in countries whose 2020 populations are greater than 30 million and whose population growth rates are greater than or equal to 0.5%.

6

Analyzing the Data of One Variable Using the Frequency Table

In the previous chapters, we have mainly studied methods of using Excel. Hereafter, we will learn data analysis methods. In the actual data analysis, it is important to summarize the data and obtain useful information. There are established methods of analysis known as descriptive statistics.

In this and the upcoming chapters, we analyze one variable. We learn about the frequency table in this chapter, and then about the parameters known as locations and scatters of the distribution of the variable in the next chapter.

6.1 Frequency Table and Histogram

Using the frequency table is one of the most basic and important methods for analyzing observed data (hereafter, observations). In the frequency table, we divide the observations into classes, and count the numbers of observations in each class (these numbers are called frequencies). The classes are determined by lower and upper limits (following Excel, observations that take the upper limit are included and those that take the lower limit are not included in that class). It is necessary to determine the values that represent the classes. These values are called "class values" and are usually chosen as the averages of the lower and upper limits,

$$\text{Class value} = (\text{lower limit} + \text{upper limit})/2$$

Suppose that the frequency is 100. However, the meaning of the value is quite different when the total numbers of observations are 1,000 and 10,000. In many cases, the fraction of frequency becomes important. We call this the fraction relative frequency (= frequency/total number of observations). The sums of the frequencies and relative frequencies from the first to the concerning class (observations smaller than or equal to the upper limit) are called the cumulative and cumulative relative frequencies, respectively. These represent the numbers and fractions of the observations that are less than or equal to the upper limits. These values become the total number of observations and 100% in the final class.

When we make a frequency table, determining the number of classes and the widths of the classes becomes important. If we make the number of classes too small, we lose a lot of information. If we make it too large, we cannot summarize the data, which is the purpose of the frequency table. Determining the number of classes looks simple but is actually a very difficult problem, and various studies have been done on this topic. Although, unfortunately, there is not an established method, we use Sturges' rule as a guide. Let n be the total number of observations and let k be the number of classes. The rule is

$$k = \log_2(n) + 1$$

(Since k is not an integer unless n is an exponentiation of 2, we choose the closest integer.) If $n = 100$, $2^6 = 64$ and $2^7 = 128$, so k becomes 7 or 8.

For reference, n and k (the integer closest to the rule's value) are given by

n	50	100	1,000	10,000	1 million	100 million	300 million
k	7	8	11	14	21	28	30

Note that k does not increase much if n increases by a large amount. If we consider the whole population of the USA (about 331 million in 2020), we just need 30 classes. To use Sturges' rule in Excel, we use the general function of calculating the logarithm and input **=LOG(*n*, 2) +1** in the proper cell. Note that Sturges' rule is just a guideline, and we do not have to set the classes as the rule suggests. Since it is not easy to understand the frequency table, we can represent the same data in a graph. We can see the distribution visually in a graph, and we can get a comprehensive view of the data. We use a bar chart to represent the frequency table graphically, and we refer to it as a histogram.

6.2 Making a Frequency Table and Histogram Using the Population Data

6.2.1 Frequency Table

Let's make a frequency table using the Countries' Population Data. Open the file of the population data. Since the current worksheet [Sheet1] contains the results obtained in the previous chapter, we insert a new worksheet and make a frequency table. Let's insert [Sheet2]. Click the down arrow for [Insert] in the [Home] tab and select [Insert Sheet] in the submenu. [Sheet2] is inserted. Copy the necessary data from Sheet1 to Sheet2. Go back to [Sheet1] by clicking [Sheet1] at the bottom of the screen. Select the range, **A1:I79**, by dragging. Click [Copy] → [Sheet2]. The sheet is changed to [Sheet2]. Move the active cell to A1. Click [Paste] and the data are pasted. We can use [Copy/Paste] to copy data among different sheets.

Next, we make a frequency table of the [Population 2020] data. Since we will use the [Population 2020] data several times, name it Population1. Select the data range for [Population 2020] (F2:F79, not including field name)→ [Formulas] tab → [Define Name] in the [Define Names] group. In the [New Name box], change the [Name] to **Population1** and click [OK].

Next, we have to set the proper classes. For that purpose, we need to know the minimum and maximum values. Input **Population 2020** in K1, **Minimum** in K2, and **Maximum** in K3. Then input **=MIN(Population1)** in L2 and **=MAX (Population1)** in L3. Among 78 countries, the smallest [Population 2020] is 4.3 million (Kuwait) and the largest is 1.42 billion (China). So, we divide the dataset into the following 6 classes:

1 million ~ 10 million
10 million ~ 50 million
50 million ~ 100 million
100 million ~ 500 million
500 million ~ 1 billion
1 billion ~ 1.5 billion.

The widths of classes are usually set to be equidistant, but we choose classes so that the characteristics of the distribution can be expressed clearly when the data values are quite different, as in this example.

In Excel, we calculate frequencies using the upper limits of the classes. The upper limit of a class is contained in that class, and the frequency of values greater than the lower limit and less than or equal to the upper limit is calculated. In the first class, the frequency is the number of values less than or equal to the first upper limit. The frequency greater than the largest upper limit is counted in [More]. (In this case, we determined the classes based on the minimum and maximum values, and so the frequency in [More] becomes zero.) From K5 to K10, input the values of the upper limits, **10000**, **50000**, **100000**, **500000**, **1000000**, and **1500000** (Fig. 6.1). (For these large numbers, we can use the scientific (exponent) notation, and input them as **1E4**, **5E4**, **1E5**, **5E5**, **1E6**, and **1.5E6**, respectively).

Next, we make the frequency table. Select the [Data] tab → [Data Analysis] in the [Analysis] group (Fig. 6.2). If [Data analysis] does not exist, insert it using the procedure explained in Chapter 1. The [Data Analysis] box appears. Click [Histogram] → [OK] (Fig. 6.3). The [Histogram] box appears. Input **Population1** in [Input Range] . Next, we input the upper limit values (**K5:K10**) in [Bin Range]. Of course, we can type K5:K10, but we can also input it using the mouse. Click the up arrow to the right of [Bin Range]. The [Histogram] box appears. Then drag K5:K10 and click the down arrow to the right of the box.

Finally, we input the output range. Click [Output Range] and input **K13** by typing or using the mouse. If you do not specify the [Output Range], the result will be written in a new worksheet (Fig. 6.4). Click [OK], and the frequencies are output to the range from K13 (Fig. 6.5).

	K	L
1	Population 2020	
2	Minimum	4303
3	Maximum	1424548
4		
5	10000	
6	50000	
7	100000	
8	500000	
9	1000000	
10	1500000	

Fig. 6.1 From K5 to K10, input the values of the upper limits.

Using these frequencies, we calculate the relative, cumulative, and relative cumulative frequencies and complete the frequency table. Move the active cell to N1 and input **Table 1 Frequency Table of 2020 Population by Countries**. From N2 to T2, input **Lower Limit**, **Upper Limit**, **Class Value**, **Frequency**, **Relative Frequency**, **Cumulative Frequency**, and **Cumulative Relative Frequency**. Input the lower limit values **1000**, **10000**, **50000**, **100000**, **500000**, and **1000000** from N3 to N8, then input the upper limit values from O3 to O8. The class value is the mean of the lower and upper limit values. Input **=(N3+O3)/2** in P3 and copy the calculated frequencies from Q3 to Q8 (we do not need the value of [More]). To obtain the relative frequencies, we calculate the total number of observations. Move the

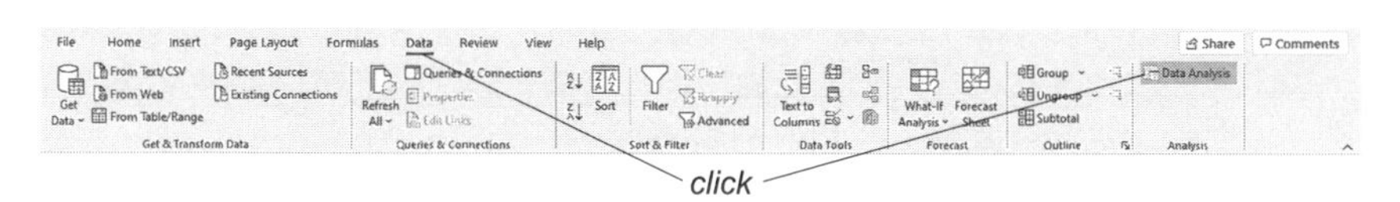

Fig. 6.2 [Data] tab → [Data Analysis] in [Analysis] group.

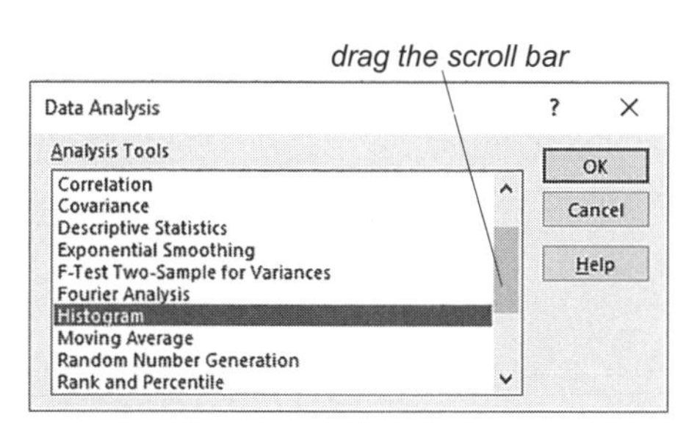

Fig. 6.3 The [Data Analysis] box appears. Click [Histogram] → [OK].

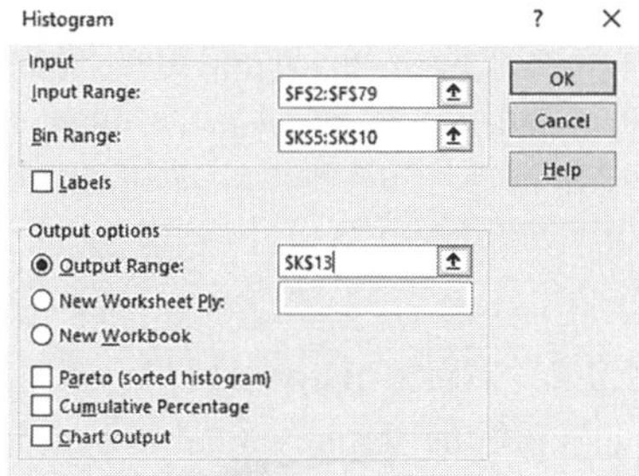

Fig. 6.4 Input **Population1** in [Input Range]. Next, we input the upper limit values (**K6:K10**) in [Bin Range]. Click [Output Range] and input **K13**.

	K	L
13	Bin	Frequency
14	10000	12
15	50000	38
16	100000	14
17	500000	12
18	1000000	0
19	1500000	2
20	More	0

Fig. 6.5 The frequencies are output from K13.

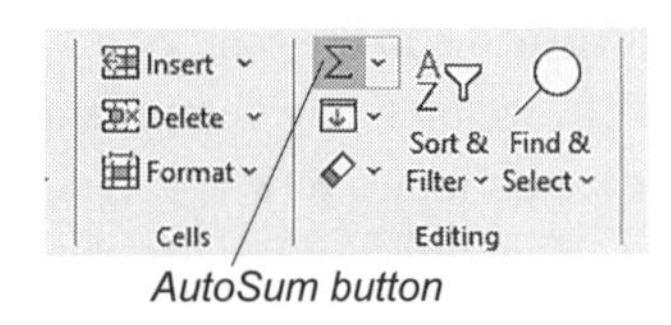

Fig. 6.6 Select [Home] tab → [AutoSum] in [Editing Group].

active cell to Q9. Select [Home] tab → [AutoSum] in [Editing Group] (Fig. 6.6). "=SUM(Q3:Q8)" is automatically input. Press the [Enter] key. The total number of observations, 78, appears in Q9. We calculate the relative frequencies. Input **=Q3/$Q$9** in R3 and copy it to R4:R8. Next, we calculate the cumulative frequencies. Since the cumulative frequency is the sum of frequencies up to that class, input **=Q3** in S3 and **=Q4+S3** in S4 and copy it to S4:S8. Finally, calculate the cumulative relative frequencies from T3 to T8.

Change the widths of the columns, alignments, and number formats (with the population data in comma format, the relative and cumulative relative frequencies to the percent format with one digit after the decimal point), and use ruled lines to complete the frequency table (Fig. 6.7).

	N	O	P	Q	R	S	T
1	Table 1 Frequency Table of 2020 Population by Countries						
2	Lower Limit	Upper Limit	Class Value	Frequency	Relative Frequency	Cumulative Frequency	Cumulative Relative Frequency
3	1,000	10,000	5,500	12	15.4%	12	15.4%
4	10,000	50,000	30,000	38	48.7%	50	64.1%
5	50,000	100,000	75,000	14	17.9%	64	82.1%
6	100,000	500,000	300,000	12	15.4%	76	97.4%
7	500,000	1,000,000	750,000	0	0.0%	76	97.4%
8	1,000,000	1,500,000	1,250,000	2	2.6%	78	100.0%
9				78			

Fig. 6.7 Change the widths of the columns, the alignments, and the number formats (put the population data in a comma format and the relative and cumulative relative frequencies in a percent format with one digit after the decimal point), and use ruled lines to complete the frequency table.

6.2.2 Histogram

Let's put the frequency table data into the form of a bar chart (histogram). For practice, we use the class values as the X (horizontal) axis. However, unlike in Chapter 3, the data are numbers. So, we make the graph as follows. First, select the frequency data (Q2:Q8) by dragging. We do not select the data for the class values at this stage. Select the [Insert] tab → [Insert Column or Bar Chart] →

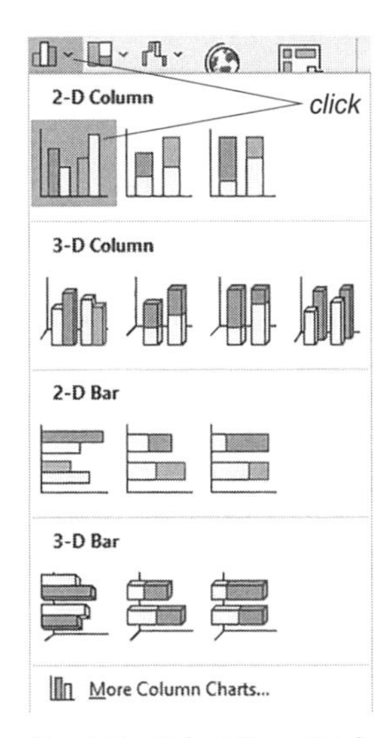

Fig. 6.8 Select [Insert] tab → [Insert Column or Bar Chart] → [Cluster Column] in [2-D Column].

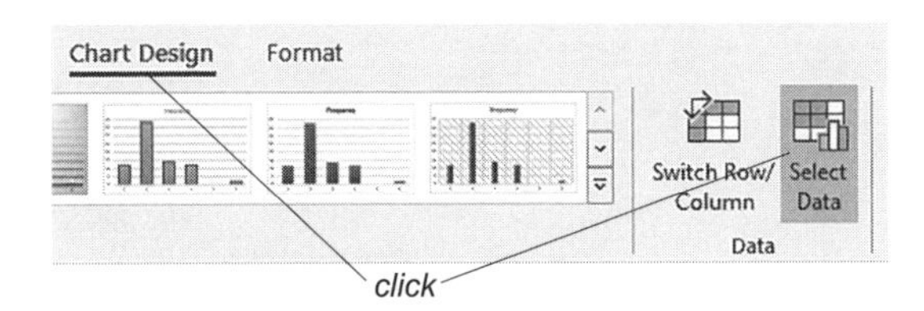

Fig. 6.9 Make the graph active and click [Design] tab → [Select Data] in the Data group.

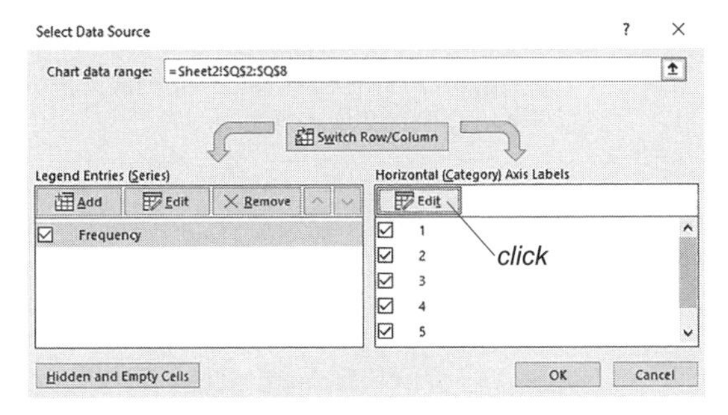

Fig. 6.10 The [Select Data Source] box appears. Click [Edit] in the [Horizontal (Category) Axis Labels].

[Cluster Column] in the [2-D Column] (Fig. 6.8). The bar chart appears.

i) Make the graph active and click the [Design] tab → [Select Data] in the Data group (Fig. 6.9).

ii) The [Select Data Source] box appears. Click [Edit] for [Horizontal (Category) Axis Labels] (Fig. 6.10).

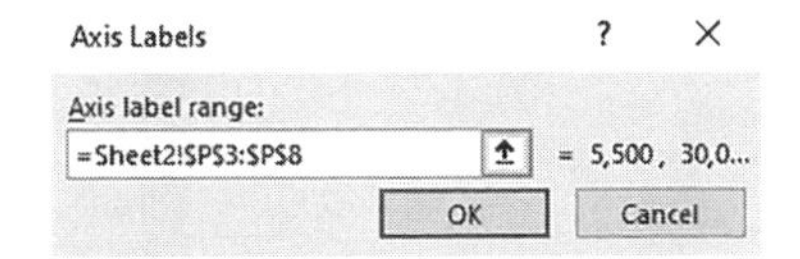

Fig. 6.11 The [Axis labels] box appears. Click [Axis label range], input the range of the class values (P3:P8) by typing or using the mouse, and click [OK].

iii) The [Axis labels] box appears. Click [Axis label range], input the range of the class values (P3:P8) by typing or using the mouse, and click [OK] (Fig. 6.11).

iv) The class values appear in [Horizontal (Category) Axis Labels]. Click [OK] (Fig. 6.12).

v) Make the title of the graph **Fig.1 Histogram of the 2020 Population of Countries**. Set the horizontal axis level to be **Population(1,000)** by selecting [Format] → [Text Box] in the [Insert Shapes] group, and set the vertical axis level to be **Frequency** (Fig. 6.13).

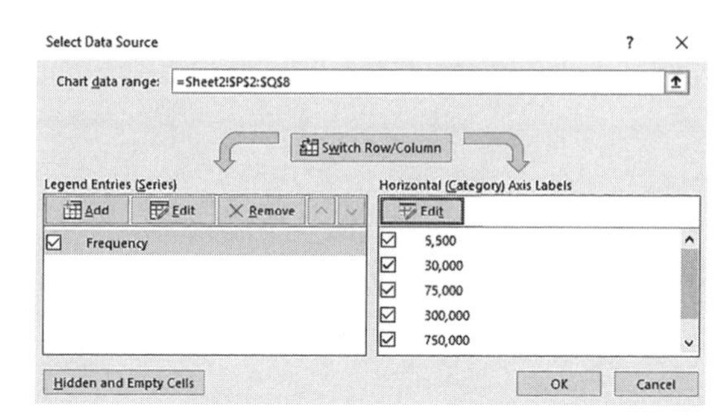

Fig. 6.12 The class values appear in [Horizontal (Category) Axis Labels]. Click [OK].

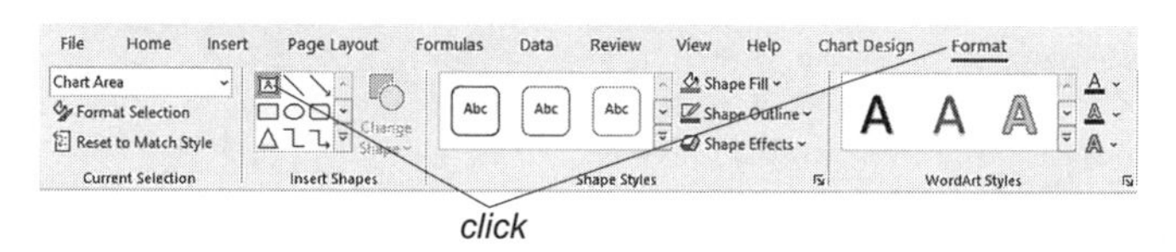

Fig. 6.13 Select [Format] →[Text Box] in the [Insert Shapes] group and insert [Text Box].

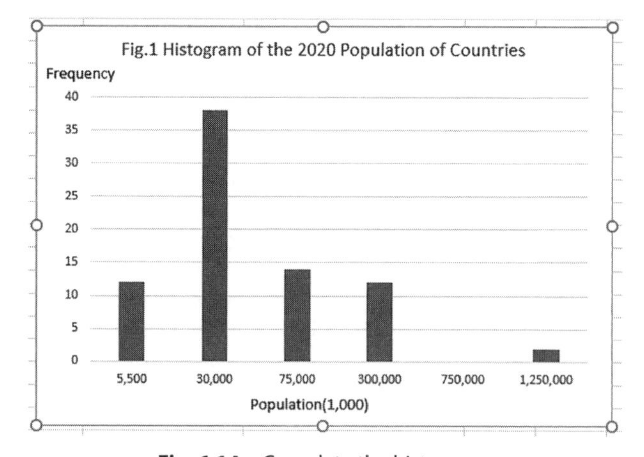

Fig. 6.14 Complete the histogram.

Complete the histogram following Chapter 3 (Fig. 6.14).

6.2.3 Cumulative Frequencies

Let's make a graph of cumulative frequencies. We use a scatter diagram (X-Y graph). The widths of the classes are quite different in this case, and the graph becomes a very strange shape if we use the original data. So, we take the logarithm of the population data for the horizontal (X) axis.

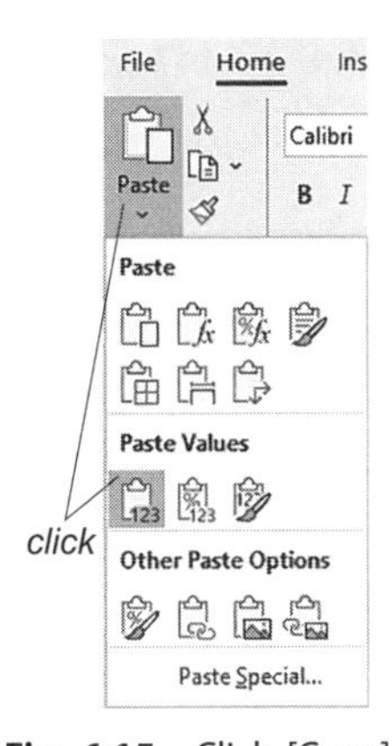

Fig. 6.15 Click [Copy] in the [Home] tab → Down arrow of [Paste] → [Paste Value].

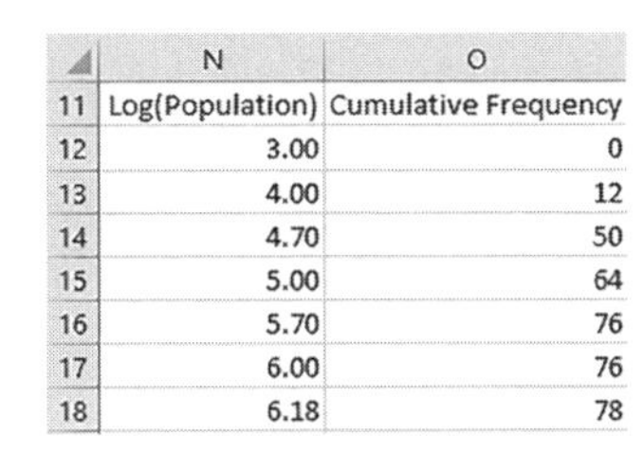

	N	O
11	Log(Population)	Cumulative Frequency
12	3.00	0
13	4.00	12
14	4.70	50
15	5.00	64
16	5.70	76
17	6.00	76
18	6.18	78

Fig. 6.16 Obtain the Log(population) and Cumulative Frequency.

Fig. 6.17 [Insert Scatter (X, Y) or Bubble Chart] in [Charts] group → [Scatter with Straight Lines].

Input **Log(Population)** in N11 and **Cumulative Frequency** in O11. Then input **=LOG10(N3)** in N12. It represents the range below the lower limit of the first class. Next, input **=LOG10(O3)** in N13 and copy it to N18. LOG10(·) is a function calculating the common logarithm, and 1 becomes 0, 10 becomes 1, 100 becomes 2,…, and the value increases by 1 each time the number of digits increases by 1. We input the cumulative frequencies in the next column. No observations exist below the lower limit of the first cell. Input **0** in O12. From O13, we copy the calculated cumulative frequencies. Since they are input as equations, we can use the "Paste Value" function. Select S3:S8 by dragging, and click [Copy] in the [Home] tab → down arrow of [Paste] → [Paste Value] (Fig. 6.15, 6.16). Finally, we make a scatter diagram of the log(Population) and cumulative frequencies. Select N11:O18 by dragging. Click [Insert] tab → [Insert Scatter (X, Y) or Bubble Chart] in [Charts] group → [Scatter with Straight Lines]. (Fig. 6.17). The scatter diagram appears. Make the title **Fig. 2 Scatter Diagram of Cumulative Frequencies of 2020 Population of Countries**. In the histogram, label the horizontal axis **Log(Population)** and label the vertical axis **Frequency** by selecting [Format] tab → [Text Box].

In the current form, the range of the horizontal axis is too large to see. So, we change the minimum and maximum values of the horizontal axis. Click a cell outside of the graph area so that the graph is not active. Then, double click a number on the horizontal axis (for example, [3.00], Fig. 6.18). The [Format Axis] box appears. Set the value of [Minimum] in [Bounds] to **2.5** and the value of [Maximum] to **6.5**, and click the [Close] button (Fig. 6.19). Complete the graph by changing the position, size, and font sizes as before (Fig. 6.20). As an exercise, make a graph of the cumulative relative frequencies.

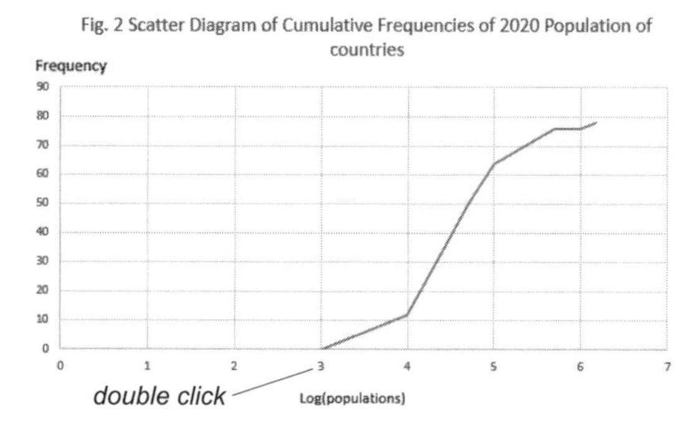

Fig. 6.18 Click a cell outside the graph area so that the graph is not active. Then, double-click a number on the horizontal axis (for example, [3.00]).

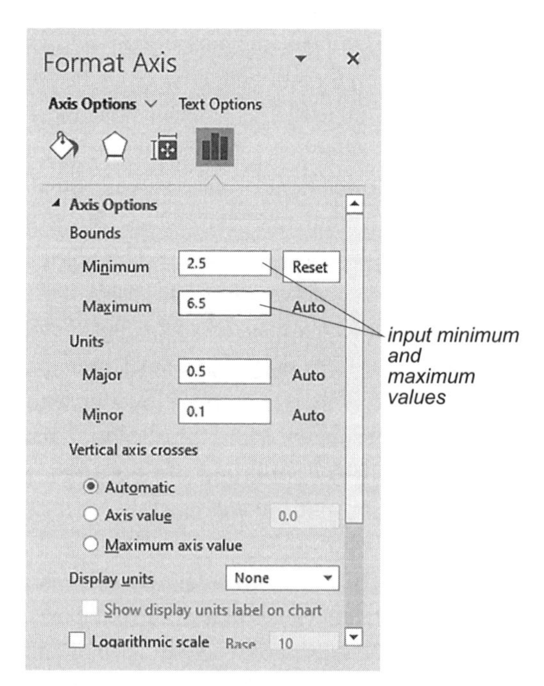

Fig. 6.19 The [Format Axis] box appears. Set the value of [Minimum] in [Bounds] to be **2.5** and the value of [Maximum] to be **6.5**, and click the [Close] button.

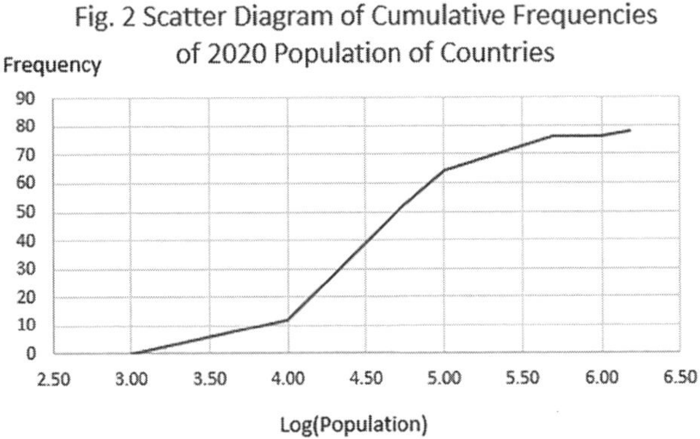

Fig. 6.20 Complete the graph by changing the position, size, and font sizes as before.

6.2.4 Lorenz Curve

The Lorenz curve is often used as an application of relative cumulative frequencies, especially in economics. The Lorenz curve is used to analyze the equality and inequality of the distributions of income or wealth. Suppose there are n people and order these people by their incomes, so that $r_1 \le r_2 \le \cdots \le r_n$,where r_i is the income of person i. We calculate the cumulative income and cumulative population from the lowest income. Then, we get

People	1	2	3	⋯	n
Cumulative income	r_1	r_1+r_2	$r_1+r_2+r_3$	⋯	$r_1+r_2+r_3+\cdots+r_n$
Cumulative population	1	2	3	⋯	n

Let the total income be $R = r_1 + r_2 + r_3 + \cdots + r_n$. Calculate the cumulative relative income by dividing by R and the cumulative relative population by dividing by the total population n. The graph in which the horizontal (X) axis is the cumulative relative population and the vertical (Y) axis is the cumulative relative income is called the Lorenz curve (Fig. 6.21).

As an extreme case, assume that the incomes of all people are the same. The graph line then becomes a diagonal line connecting the lower left and upper right corners. In another extreme case, let the income of person 1 to person $n - 1$ is zero and person n monopolizes all of the income. The graph line sticks to the lower and right sides of the frame. As the degree of inequality increases, the graph departs from the diagonal line and moves toward the lower right. This means that as the distance from the diagonal line becomes larger, the inequality of the distribution of income becomes larger. The area between the graph and the diagonal line represents the inequality of the distribution. The value that doubles the area is known

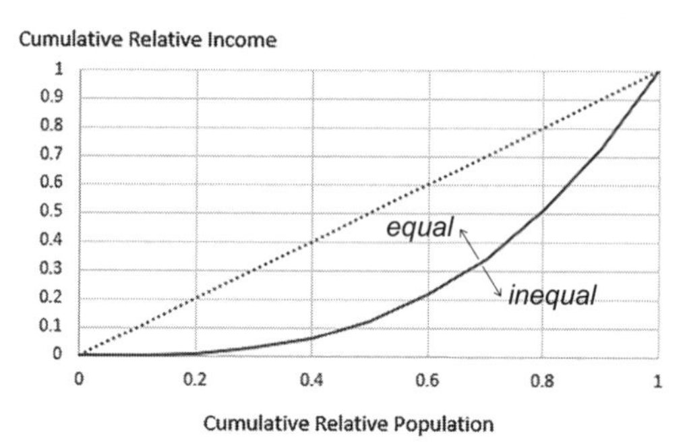

Fig. 6.21 Lorenz curve.

as the Gini index, which is widely used to express the inequality of an income distribution.

Let's make the Lorenz curve using the population data and analyze the inequality of the international distribution of income. Here, we do not consider the income distribution within a country. Instead, we assume that all people in the country have the same income represented by [GDP per capita]. Copy the [GDP per capita] data and [Population 2020] to the range beginning with A85 including the field names. Sort the copied data by GDP per capita. Move the active cell to the range of the new List, click the [Data] tab → [Sort] in [Sort & Filter]. Select [GDP per capita] in [Sort by], set [Order] to [Smallest to Largest], and click [OK].

Now, we calculate the GDPs of countries by multiplying GDP per capita by the population. The population is given in units of 1,000, and the GDP per capita is given in dollars. The GDP values become too large using this unit, so instead we calculate GDP using the unit of one billion dollars. Input **GDP** in C85 and **=A86*B86/1E6** in C86. 1E6 is 1,000,000 given in scientific style. Copy C86 to all of the countries.

Next, we calculate the cumulative population and cumulative GDP. Input **Cumulative Population** and **Cumulative GDP** in D85 and E85, respectively. We calculate the population in units of one million. Input **=B86/1000** in D86 and **=D86+B87/1000** in D87 and copy this to all countries. Then, we calculate the cumulative GDP. Input **=C86** in E86 and **=E86+C87** in E87 and copy it to column E.

Next, we calculate the cumulative relative population and GDP. Input **Cumulative Relative Population** and **Cumulative Relative GDP** in F85 and G85, respectively. Since the totals are calculated in D163 and E163, input **=D86/D$163** in F86 and copy it to Columns F and G. We get the cumulative relative population and GDP (Fig. 6.22).

Finally, we complete the Lorenz curve by making the scatter diagram (X-Y

	A	B	C	D	E	F	G
85	GDP per capita	Population 2020	GDP	Cumulative Population	Cumulative GDP	Cumulative Relative Population	Cumulative Relative GDP
86	338	20,284	7	20	7	0.29%	0.01%
87	378	24,075	9	44	16	0.64%	0.02%
88	426	32,309	14	77	30	1.10%	0.04%
89	450	27,691	12	104	42	1.50%	0.05%
90	463	89,505	41	194	84	2.78%	0.11%
91	642	20,903	13	215	97	3.08%	0.12%

Fig. 6.22 Sort the data by GDP per capita and obtain the cumulative relative population and GDP.

graph). Select the ranges of the [Cumulative Relative Population] and [Cumulative Relative GDP] including the field names by dragging. Click [Insert] tab → [Insert Scatter (X,Y) or Bubble Chart] → [Scatter with Straight Lines]. Make the title **Fig. 3 Lorenz Curve Based on Population Data**. Label the horizontal and vertical axes as **Cumulative Relative Population** and **Cumulative Relative GDP** by selecting [Format] → [Text Box]. The graph of the Lorenz curve appears. Complete it by adjusting the graph size, position, axis scales, and font sizes (Fig. 6.23).

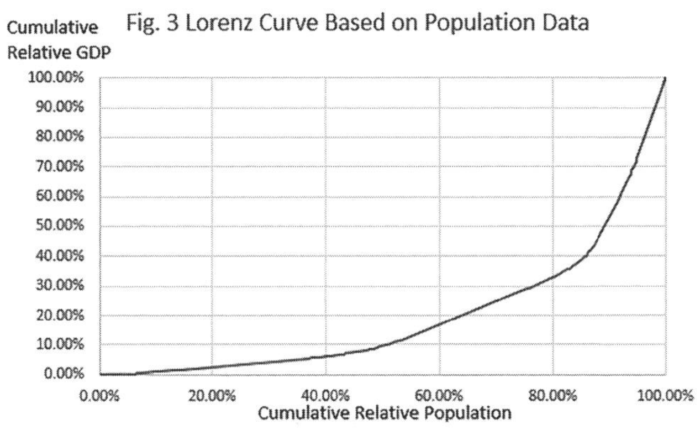

Fig. 6.23 Complete the Lorenz curve.

6.3 Exercises Using the Population Data of Countries

1. Make frequency tables of [Population Density] and [Population Growth Rate], and make histograms of the frequencies and graphs of the cumulative frequencies.
2. Make the Lorenz curve using [Population 2070] and [GDP per capita].

7

Locations and Scatter Scales of the Distribution of the Variable

Analysis using the frequency table has an advantage in that we can judge the characteristics of the distribution visually and intuitively from the histogram. However, it is difficult to treat the frequency table mathematically, and the shape of the histogram will change if we change the classes. Locations and scatter scales are indicators that can provide summaries of the distribution. Locations and scatter scales are calculated objectively and strictly, and theoretical treatments of these indicators are easy.

Of course, how well the locations and scatter scales represent the distribution depends on the distribution. The characteristics of the distribution can be obtained by analyzing the frequency table and histogram, though analyzing only one of these is not enough, and they are complementary to each other.

7.1 Locations of the Distribution

The locations of the distribution are given by several different indicators. The indicators representing and summarizing the data are called statistics. Here, we learn about the important statistical measures of the average (mean), weighted average, median, and percentile.

7.1.1 Average (Mean)

a. Average (Mean)

The average (or mean) is more precisely known as the arithmetic mean, but we can use the term “average” or “mean” to refer to the arithmetic mean. In accordance with how the function is named in Excel, we use the term “average” in this chapter. The average is the most widely used indicator representing the locations of the distribution. The average $\bar{x}$ is calculated as the total sum of the observations divided by the number of observations:

$$\bar{x} = \frac{x_1 + x_2 + \cdots + x_n}{n} = \sum x_i / n \tag{7.1}$$

where $\sum$ is the operator signifying the summation of all values from x_1 to x_n. In Excel we can calculate the average using the function **AVERAGE(Data Range)**.

b. Weighted Average

Let's calculate the GDP per capita of all listed countries from the Countries' Population Data we are using. The population differs among the countries; we cannot just calculate the GDP per capita from the simple average. In this case, we use the weighted average depending on each country's population. The weighted average $\bar{x}_w$ is obtained by

$$\bar{x}_w = \sum w_i x_i, \ \sum w_i = 1, \ w_i \geq 0 \tag{7.2}$$

$w_1, w_2, \ldots, w_n$ are the weights and relative populations of the countries, and the sum of all of them equals 1. If all weights are the same and $1/n$, the weighted average is the same as the simple average. We use the weighted average when we calculate the average from the frequency table.

7.1.2 Median

Let's sort the observations from the smallest to largest. The median x_M is the center or 50% value and is defined as

(7.3)

$$x_M = \begin{cases} \text{value of the}(n+1)/2\text{th observation if } n \text{ is odd} \\ \text{value of } \{n/2\text{th observation } + (n/2+1)\text{th observation}\}/2 \text{ if } n \text{ is even} \end{cases}$$

If some observations are much larger or smaller than the other values and the distribution is fat-tailed, the average is not a good indicator to represent the location. In this case, the median is a better indicator. In Excel, we can obtain the median using **MEDIAN(Data Range)**.

7.1.3 Percentile

Median can be generalized to the p-th percentile, p% value from the smallest, x_p. Let $m = n{\cdot}p$ and m^* be the smallest integer that is larger than or equal to m, and generalizing (7.3), define

(7.4)

$$x_p = \begin{cases} \text{value of } x_{m^*} \text{ if } m \text{ has a decimal part} \\ \text{value of } \{m^*\text{th observation } + (m^*+1)\text{th observation}\}/2 \text{ if } m \text{ is an integer.} \end{cases}$$

By this definition, the p-th percentile from the smallest and $(100-p)$ th percentile from the largest become the same value. The median is the 50th percentile and is included within this general definition.

In Excel, the p-th percentile calculated using **PERCENTILE(Data Range, p%)**.

However, it uses linear interpolation, and the value may be different from that obtained using definition (7.4), which is a natural generalization of (7.3).

In data analysis, percentiles that are very widely used are quartiles. They are the 25th, 50th, and 75th percentiles, and they divide the data into quarters (25%), which, beginning with the smallest, are called the first, second (median), and third quartiles.

In Excel, the quartiles are calculated using **Quartile(Data Range, Quart)**, Quart takes integers from 0 to 4; 0: minimum, 1: first quartile, 2: second quartile (median), 3: third quartile, and 4: maximum.

7.2 Scatter Scales

Let's consider the distributions A and B in Fig. 7.1. In these distributions, the mean and median are the same, but B is more scattered (widely spread) than A. Scatter scales represent the shapes of distributions, i.e., how they are scattered around the defined locations. Here, I explain the range, interquartile range, variance, and standard deviation.

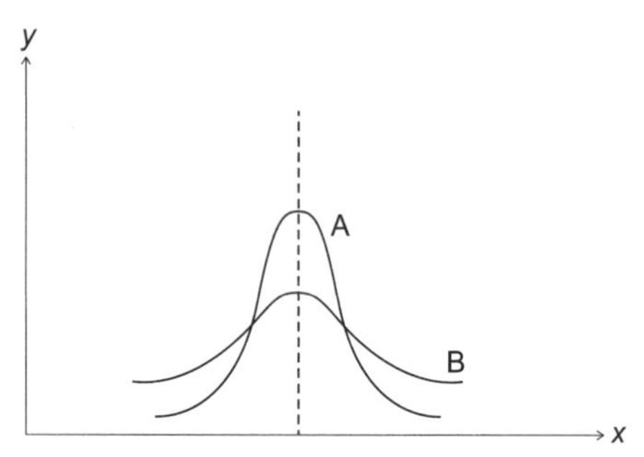

Fig. 7.1 Two different distributions.

7.2.1 Range

The range is the difference between the maximum and minimum observations. The range gives rough information about the spread of the distribution since it is obtained just from the maximum and minimum. It becomes a large value if there exist extremely large or small observations.

7.2.2 Interquartile Range

The interquartile range is given by

(7.5) 75th percentile (third quartile) – 25th percentile (first quartile)

Unlike the range, it is not affected by extreme observation values. However, the problem is that all observations are not used; that is, the observations larger than the 75th percentile or smaller than the 25th percentile do not affect the interquartile range.

7.2.3 Variance and Standard Deviations

As scatter scales, the indicators used most are based on the deviation, which expresses the distance of the observation from the mean and is calculated by the observation x_i − the average (i.e., $x_i - \bar{x}$).

The deviations take both positive and negative values. If we simply sum them up, we always get zero. To remove the effects of signs, we square them, sum them up, and divide by the number of observations n. This value is called the variance. (To calculate the variance, we also divide the sum of squared deviations by $n - 1$. At the end of this section, we will learn the difference between these two methods (dividing by n or $n - 1$).)

The variance S^2 is obtained by

$$S^2 = \sum (x_i - \bar{x})^2 / n \tag{7.6}$$

Since the variance involves squaring the deviations, we cannot directly compare it to the observations. For example, the unit changes with squaring. So, we take the square root of the variance and get S, which is called the standard deviation, and we use S in the analysis. We will learn the meaning of the variance and standard deviation in later chapters.

In Excel, the variance is calculated using **VARP(Data Range)**, and the standard deviations are calculated using **STDEVP(Data Range)**.

Dividing the Sum of Squared Deviations by *n* or *n* − 1

The set of all elements that we are interested in is called the *population*. (It is not a number of people. In the dataset, a population is the number of people in a country (i.e., we use the Countries' Population Data), and this is the meaning of the term that we use in practices and exercises. We use *population* (in italics) when we use it to mean the set of all elements.) However, a *population* usually contains many elements, and it is very difficult to do a complete survey of the *population* (*population* survey). In such cases, we select a part of the *population* and survey it. The part selected from the *population* is called the sample, and we survey the selected sample (i.e., we perform a sample survey). For example, media companies conduct sample surveys selecting a few thousand people almost every week.

In statistics, we divide the sum of the squared deviations by n when we do a *population* survey and by $n - 1$ when we do a sample survey. As we will learn in detail in Chapters 10 and 11, we can analyze the data easily if we divide by $n - 1$

for the sample survey.

In Excel, the variance and standard deviation are calculated using:

Population (dividing by n): **VARP, STDEVP**

Sample (dividing by $n - 1$): **VAR, STDEV**

7.3 Calculation of Locations and Scatter Scales from the Frequency Table

Locations and scatter scales can be calculated from the frequency table. (Of course, they are approximations because of the intervals of the classes.) Here, we calculate the average, median, p-th percentiles, variance, and standard deviation.

7.3.1 Average

In the frequency table, the class values represent the classes, and we calculate the average assuming that observations in the class take the class value. This is equivalent to obtaining the weighted average of the class values where the weights are the relative frequencies of the classes. Let $\bar{x}^*$ be the average calculated from the frequency table. Then

$$\bar{x}^* = r_1 m_1 + r_2 m_2 + \cdots + r_k m_k = \sum r_i m_i \tag{7.7}$$

where r_i is the relative frequency, m_i is the class value of class i, and k is the number of classes.

7.3.2 Median and p-th Percentiles

To obtain the median from the frequency table, we first determine the class that contains the median. The median is the 50th percentile. We need to find the class that the cumulative relative frequency of the previous class is less than 50%, and the cumulative relative frequency including that class is greater than or equal to 50%. Since the cumulative relative frequency monotonically increases, there usually exists only one class that satisfies these conditions. Let x_L be the lower limit, x_U be the upper limit of that class, R_0 be the cumulative relative frequency of the previous class, and R_1 be the cumulative frequency including that class. R_0 and R_1 satisfy $R_0 < 50\% \leq R_1$. The median is close to x_L if R_0 is close to 50% and close to x_U if R_1 is close to 50%. So, we divide the intervals proportionally, and the median x_M^* is obtained by

$$x_M^* = x_L + (x_U - x_L) \cdot \left\{ \frac{50\% - R_0}{R_1 - R_0} \right\} \tag{7.8}$$

This corresponds to determining the 50% point using the cumulative relative fre-

quency graph.
To generalize the median, the p-th percentile x_p^* is obtained by finding the class satisfying $R_0 < p\% \leq R_1$ and dividing the interval based on the values of R_0 and R_1, and

$$x_M^* = x_L + (x_U - x_L) \cdot \left\{ \frac{p\% - R_0}{R_1 - R_0} \right\} \tag{7.9}$$

As the median case, x_L and x_U are the lower and upper limits of the class, respectively. R_0 is the cumulative relative frequency of the previous class, and R_1 is the cumulative frequency of the class.

7.3.3 Variance and Standard Deviation

As the average case, the variance is calculated assuming all observations in the class take the class value. Therefore, the variance S^{*2} becomes the weighted average of the square of (class value − $\bar{x}^*$), and is obtained by

$$\begin{aligned} S^{*2} &= r_1(m_1 - \bar{x}^*)^2 + r_2(m_2 - \bar{x}^*)^2 + \cdots + r_k(m_k - \bar{x}^*)^2 \\ &= \sum r_i(m_i - \bar{x}^*)^2 \end{aligned} \tag{7.10}$$

where $\bar{x}^*$ is the average calculated from the frequency table, and m_i and r_i are the class value and relative frequency of the class i, respectively. We can obtain the standard deviation S^* by taking the square root of the variance. (Hereafter, we call (class value − $\bar{x}^*$) "the deviation of the class value" or simply "the deviation.")

7.4 Calculation of Locations and Scatter Scales Using the Population Data

7.4.1 Calculation from the Original Data

Here, we calculate locations and scatter scales using the Countries' Population Data. The current worksheet [Sheet2] includes a frequency table and various graphs, and it is difficult to see. So, let's insert a new worksheet, [Sheet3]. Click the down arrow for [Insert] → [Insert Sheet] in the [Home] tab, and insert [Sheet3]. Copy the data of [Population 2020] to A1:A79, including the field name, in [Sheet3].

As locations, we calculate the average, 25th percentile (first quartile), 50th percentile (median), and 75th percentile (third quartile). As scatter scales, we calculate the range, interquartile range, variance, and standard deviation. Select A2:A79 by dragging and click the [Formulas] tab → [Define Name] in the [Define Names] group, and name the range **Population2**.

Now, we calculate the average and other statistics, input **From Original Data** in E1 and input **Average**, **25th Percentile**, **Median**, **75th Percentile**, **Range**, **Interquartile Range**, **Variance**, and **Standard Deviation** in E2:E9. Move the active cell to F2, input **=AVERAGE(Population2)**, and calculate the average.

Next, we obtain percentiles. First, we sort the data. Move the active cell to the List (field name +data). Select [Data] tab → [Sort] in the [Sort & Filter] group → [Smallest to Largest] sorting order→ [OK]. The data is sorted from the smallest to largest values. We obtain three percentiles. Since 78 × 25% = 19.5, 78 × 50% = 39, and 78 × 75% = 58.5, following definitions (7.3) and (7.4):

25th percentile: 20th observation

Median: average of 39th and 40th observations

75th percentile: 59th observation

The 20th observation is A21, the 39th observation is A40, the 40th observation is A41, and the 59th observation is A60. Input **=A21** in F3, **=(A40+A41)/2** in F4, and **=A60** in F5, and calculate the quartiles.

The range is the maximum – the minimum. Input **=A79–A2** in F6. The interquartile range is the 75th percentile – the 25th percentile. Input **=F5–F3** in F7.

Input **Deviation** in B1 and **Squared Deviation** in C1. The deviation is an observation minus the average. Input **=A2–$F$2** in B2 and copy it to B79. Input **=B2^2** in C2, copy it to C79, and calculate the squared deviations. As they become large numbers, make the format [Scientific]. Calculate the sum of the squared deviations in C80 using [AutoSum] in the [Home] Tab. Input **C80/COUNT(Population2)** in F8. **COUNT** is the function that counts the number of cells that contain the data in the range, and using it we can obtain *n*. Input **=SQRT(F8)** in F9 and calculate the standard deviation. (**SQRT** calculates the square root. We can get the answer if we input **=F8^0.5**, practically, the difference between the two methods will be very small. However, since specialized functions can usually calculate more accurately and quickly than general functions, we use the specialized functions when they are available.)

As we have already learned, Excel has functions for calculating the average, quartiles, variance, and standard deviation, as follows:

25th percentile: **=QUARTILE(Population2,1)**

Median: **=MEDIAN(Population2)**

75th percentile: **=QUARTILE(Population2,3)**

Range: **=MAX(Population2)–MIN(Population2)**

Variance: **=VARP(Population2)**

Standard deviation: **=STDEVP(Population2)**

Input **By Functions** in G1 and input these functions in G3:G9 and compare the

	E	F	G
1	From Original Data		By Functions
2	Average	89,443	
3	25th Percentile	17,336	17,422
4	Median	32589	32,589
5	75th Percentile	69,411	68,957
6	Range	1,420,245	1,420,245
7	Interquartile Range	52,075	51,535
8	Variance	4.9420E+10	4.9420E+10
9	Standard Deviation	222,305	222,305

Fig. 7.2 Locations and scatter scales.

results (Fig. 7.2). The values of the 25th and 75th percentiles are a little different due to the difference in the definitions, but the results are the same for the other values. Hereafter, we will use functions, and we will not have to do complicated computations anymore.

7.4.2 Calculation from the Frequency Table

We calculate the locations and scatter scales of the data from the frequency table. Copy the "Table 1 Frequency Table of 2020 Population of Countries" in [Sheet2] to the range beginning with E11 of [Sheet3], including the title and item names. The frequency table in [Sheet2] uses equations. Use the paste value (when you paste the data, click the downward arrow for [Paste] → [Values(V)] in [Paste Values]).

First, we calculate the average. The average is the sum of the products of the class values and relative frequencies. Input **C. Value * R. Freq**. in L12. Input **=G13*I13** in L13 and copy it to L18. Using the [AutoSum] and summing them up, calculate the average in L19 (Fig. 7.3). Next, we calculate the variance. The variance is obtained from the sum of the products of the squared deviations of the class values and relative frequencies. Input **Sq. Dev. * R. Freq**. in M12. Input **=(G13–$L$9)^2*I13** in M13, copy it to M18, and calculate the sum in M19 (Fig. 7.4). Next, we calculate the quartiles. The classes that contain the quartiles are

	H	I	J	K	L
11	ı of Countries				
12	Frequency	Relative Frequency	Cumulative Frequency	Cumurative Relative Frequency	C. Value * R. Freq
13	12	15.4%	12	15.4%	846
14	38	48.7%	50	64.1%	14,615
15	14	17.9%	64	82.1%	13,462
16	12	15.4%	76	97.4%	46,154
17	0	0.0%	76	97.4%	0
18	2	2.6%	78	100.0%	32,051
19					107,128

Fig. 7.3 Calculation of the average from the frequency table.

Quartile	Lower Limit x_L	Upper Limit x_U
25%	10,000	50,000
50%	10,000	50,000
75%	50,000	100,000

Input **Quartile**, **25%**, **50%**, **75%** in E21:E24. Input the following equations from F22 to F24 and calculate the quartiles.

=E14+(F14–E14)*(E22–K13)/(K14–K13)
=E14+(F14–E14)*(E23–K13)/(K14–K13)
=E15+(F15–E15)*(E24–K14)/(K15–K14)

	L	M
11		
12	C. Value * R. Freq	Sq. Dev. * R. Freq.
13	846	1.5890E+09
14	14,615	2.8981E+09
15	13,462	1.8527E+08
16	46,154	5.7230E+09
17	0	0.0000E+00
18	32,051	3.3491E+10
19	107,128	4.3887E+10

Fig. 7.4 Calculation of the variance from the frequency table.

Input **From Frequency Table** in H1. We have calculated the average, variance, and quartiles. Copy them to the range beginning with H2. Since equations are used, copy and paste them using [Values (V)] of [Paste]. Calculate the range from the largest upper limit minus the smallest lower limit, calculate the interquartile range from the 25th and 75th percentiles, and compute the standard deviation using the SQRT function. The values obtained from the frequency table are approximations, and there will be some errors (Fig. 7.5).

	E	F	G	H
1	From Original Data		By Functions	From Frequency Table
2	Average	89,443		107,128
3	25th Percentile	17,336	17,422	17,895
4	Median	32589	32,589	38,421
5	75th Percentile	69,411	68,957	80,357
6	Range	1,420,245	1,420,245	1499000
7	Interquartile Range	52,075	51,535	62,462
8	Variance	4.9420E+10	4.9420E+10	4.3887E+10
9	Standard Deviation	222,305	222,305	209491.1331

Fig. 7.5 Results of location and scatter scale calculations.

7.5 Exercises Using the Population Data

Calculate the average, quartiles, range, interquartile range, variance, and standard deviation of [Population Growth Rate].

1. From the original data not using functions (except AVERAGE).
2. With Excel functions.
3. From the frequency table made in the exercises in Chapter 6.

Compare the results.

8

Analysis of Two-Dimensional Data

In Chapters 6 and 7, we have learned about the analysis of one variable. In this chapter, we learn the methods of analyzing and summarizing two-dimensional data. With two-dimensional data, we get the values of two variables (x_i, y_i) at the same time for the subject i, such as the height and weight of person i. When working with two-dimensional data, the relationship between the two variables becomes very important. In this chapter, we learn how to perform such analysis using scatter diagrams, contingency tables, and correlation coefficients.

8.1 Scatter Diagram and Contingency Table

8.1.1 Scatter Diagram

With two-dimensional data, we get a pair of two variables (x_i, y_i) at the same time for the subject i. When these two variables are numerical variables, we can make a graph and analyze the relations of the two variables. For example, for the graph shown in Fig. 8.1(a), we can consider that there is an obvious relationship between the two variables. However, for the graph shown in Fig. 8.1 (b), we cannot conclude that the two variables are related. This graph is called a scatter diagram (or a scatter plot or scatter chart in Excel). When we make a scatter diagram, we assign a cause or explanatory variable to the X-axis and a result or explained variable to the Y-axis. We refer to this data as quantitative data.

8.1.2 Contingency Table

When the data consist of qualitative variables, such as gender (male, female) and academic background (graduating high school, university (undergraduate), graduate school) for which we know the categories or states to which the variables belong, the data is called qualitative data as mentioned in Chapter 4. For qualitative data, we cannot make a scatter diagram. In this case, we make a contingency table based on the categories that the variables can take and count the numbers of

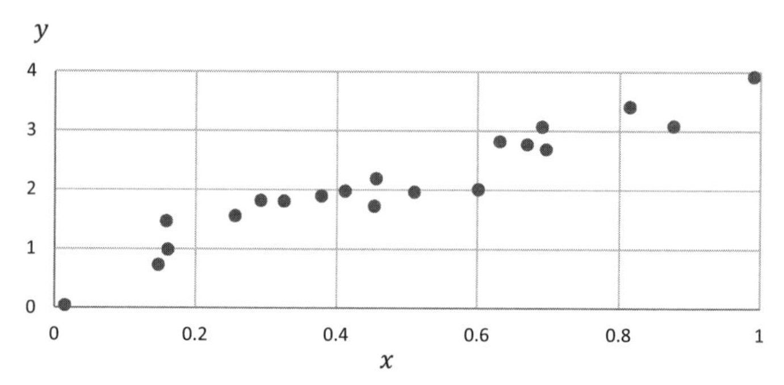

(a) We can conclude that there is an obvious relationship between two variables.

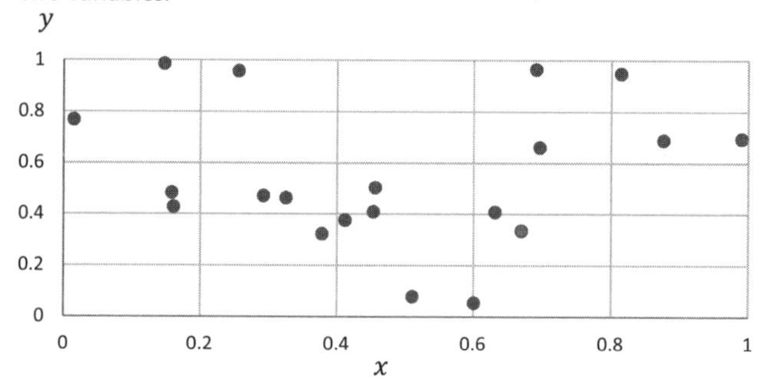

(b) We cannot conclude that the two variables are related.

Fig. 8.1 Relations of two variables.

observations in the crosses. The table is called a contingency table (or cross-tabulation table). Usually, we put values that the category x takes in the left column and the categories for y in the head row of the table. If x takes s different categories and y takes t different categories, the number of crosses (cells) become $s \times t$ and the table is called the $s \times t$ contingency table. If one or two variables are quantitative data, we divide them into classes. If we divide the range of x into s classes and that of y into t classes, the number of cells becomes $s \cdot t$. So, we have to be careful not to make s and t too large. Unfortunately, there is no predetermined rule for determining the number of classes, and we have to choose classes based on the number of observations and the purpose of the analysis.

To determine the relationship between the variables, we calculate the relative frequencies. However, unlike in the one variable case, we can consider three different relative frequencies (ratios). These include the ratio of the number of observations whose frequency is being determined to (i) the total number of observations (grand total); (ii) the total of the row to which the cell belongs (row total); (iii) the total of the column frequencies (column total). The totals determined in

(ii) and (iii) are called marginal frequencies; however, we call them "totals" in this chapter to be consistent with the output of Excel. They are used depending on the purpose of the analysis. However, the ratio to the row total is generalized because we take the categories of the explanatory or cause variables in the column.

8.2 Making a Scatter Diagram and Contingency Table from the Population Data

8.2.1 Scatter Diagram

Let's make a scatter diagram of the [GDP per capita] and [Population Growth Rate] using the Countries' Population Data. In this case, our purpose is to analyze the effects of [GDP per capita] on [Population Growth Rate]. We take [GDP per capita] as the X-axis and [Population Growth Rate] as the Y-axis. Click the down arrow for [Insert] on the [Home] tab → [Insert Sheet] and insert [Sheet4]. Copy the [GDP per capita] data in [Sheet1] to A1:A79 of [Sheet4] including the field name. The maximum value of [GDP per capita] is more than 100 times larger than the minimum, and it is not suitable to use these values in the graph, so we take the common logarithm. Input **Log of GDP** in B1 and **=LOG10(A2)** in B2 and copy it to B3:B79. (Although we can change the scales of the X and Y axes to log scales in Excel, we use the [Log of GDP] in later practices.) Next, copy the [Population Growth Rate] data to C1:C79. They are calculated using equations. Use [Paste Values] as before.

We make a scatter diagram. Select the data range B1:C79 by dragging, and click [Insert] tab →[Scatter] in the [Chart] group → [Scatter (Markers only)] (Fig. 8.2). Make the title **Fig 1. Scatter Diagram of GDP per Capita and Population Growth Rate**. Make the graph active, click [Format] → [Text box], insert the text box, and label the X-axis **Log of GDP per Capita** and the Y-axis **Population Growth Rate**. Click a cell outside of the graph to make the graph inactive, and double-click a number on the X-axis. [Format Axis] appears. Change [Minimum] to **2.0**, [Maximum] to **5.0**, and click the [Close] button (Fig. 8.3). Complete the graph by changing the position, size, and font sizes. There is a clear tendency for the population growth rate to decrease as the GDP per capita increases (Fig. 8.4).

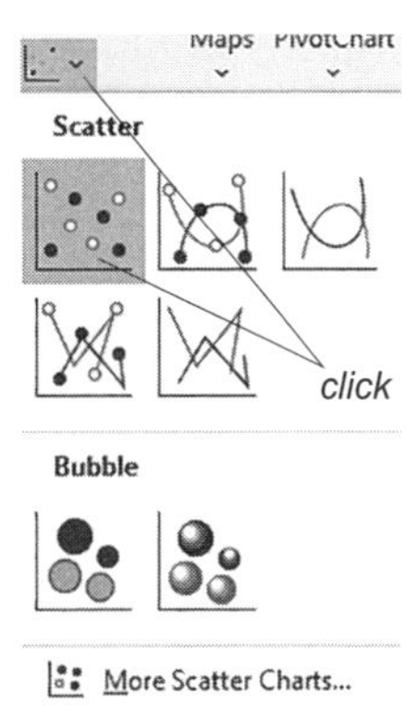

Fig. 8.2 Click [Insert] tab → [Scatter] in [Chart] group → [Scatter (Markers only)].

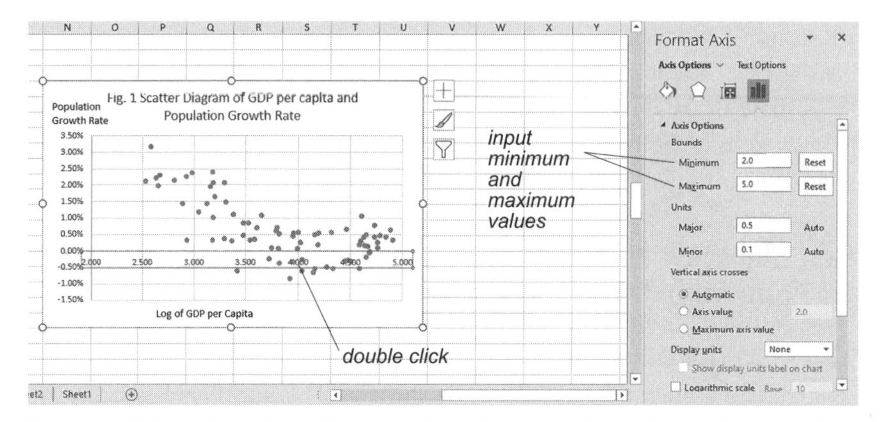

Fig. 8.3 Click a cell outside of the graph to make the graph inactive, and double-click a number on the X-axis. [Format Axis] appears. Change [Minimum] to **2.0**, [Maximum] to **5.0**, and click the [Close] button.

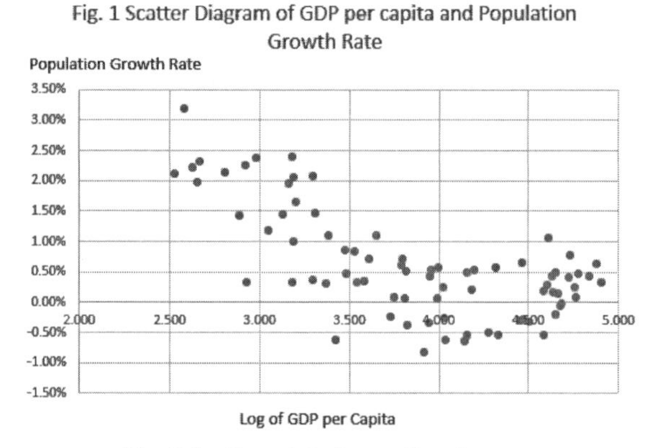

Fig. 8.4 Complete the scatter diagram.

8.2.2 Contingency Table

We make a contingency table using [PivotTable]. [PivotTable] is a feature of Excel that allows us to do complicated analyses of the data. Both [GDP per capita] and [Population Growth Rate] are quantitative data, so we divide them into classes using the **IF** function. We divide [GDP per capita] into two classes with \$16,000 as the dividing line. We refer to countries with a GDP per capita more than or equal to \$16,000 as [High Income Countries] and those with a GDP per capita less than \$16,000 as [Low Income Countries]. Input **Income** in D1 and **=IF(A2<16000,"Low","High")** in D2 and copy D2 to D3:D79 (Fig. 8.5). For the **IF** function, we put the condition as the first argument (in this case, less than \$16,000), the operation when the condition is true as the second argument (display Low), and the operation when the condition is false (display High) as the third argument. As the second and third arguments, we can use equations, functions, numbers, and characters. When we use characters, surround them by double quotations (" ").

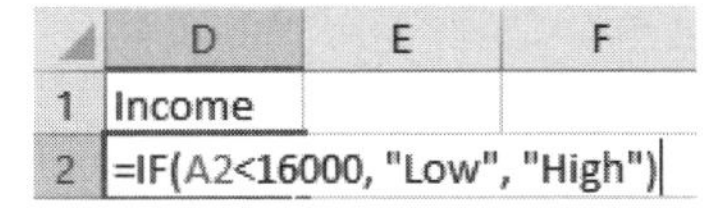

	D	E	F
1	Income		
2	=IF(A2<16000, "Low", "High")		

Fig. 8.5 Input **=IF(A2<16000, "Low", "High")** in D2.

Next, we classify the [Population Growth Rate] into three classes: high-growth-rate countries with growth rates more than or equal to 1.0%, middle-growth-rate countries with growth rates between 1.0% and 0.5%, and low-growth-rate countries with growth rates less than 0.5%. Input **Population Growth** in E1 and **=IF(C2<0.005, "Low", IF(C2< 0.01,"Middle", "High"))** in E2, and copy E2 to the data range (Fig. 8.6).

	E	F	G	H	I
1	Population Growth				
2	=IF(C2<0.005, "Low", IF(C2<0.01, "Middle", "High"))				

Fig. 8.6 Input **=IF(C2<0.005,"Low", IF(C2< 0.01,"Middle", "High"))** in E2.

To easily use the Pivot Table, input **Frequency** in F1 and **1** in F2:F79. Move the active cell to a cell in the List (A1:F79). Click [Insert] tab → [Pivot Table] in the [Tables] group (Fig. 8.7). The [Create PivotTable] box appears. So,

i) Check that [Table/Range] is correct (if not, correct it to A1:F79).
ii) Click [Existing Worksheet] in [Choose where you want the PivotTable report to be placed] and input **A86** in [Location] by typing or using the mouse.

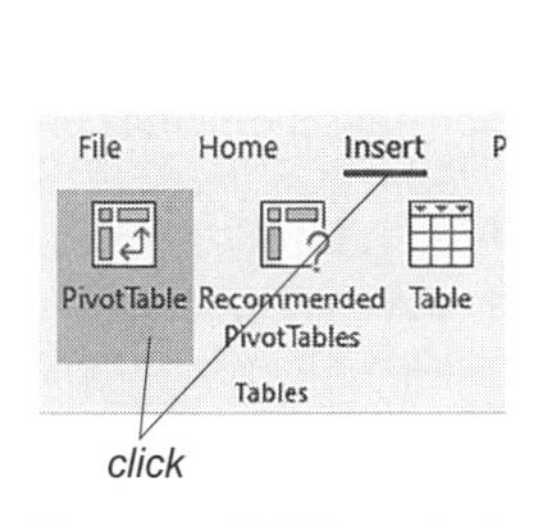

Fig. 8.7 Click [Insert] tab → [PivotTable] in [Tables] group.

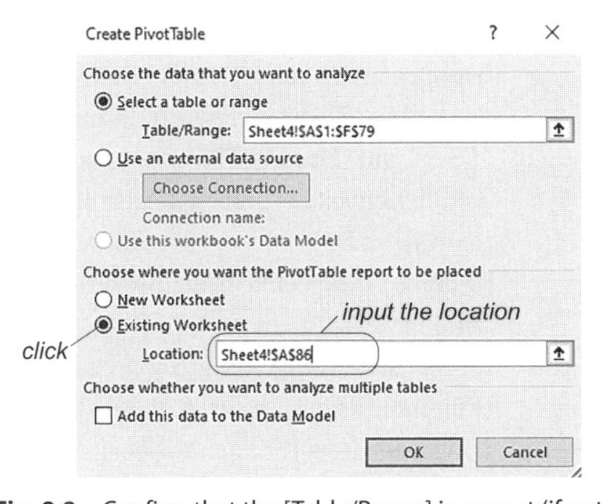

Fig. 8.8 Confirm that the [Table/Range] is correct (if not, correct it to A1:F79). Click [Existing Worksheet] in [Choose where you want the PivotTable report to be placed], input **A86** in [Location], and click [OK].

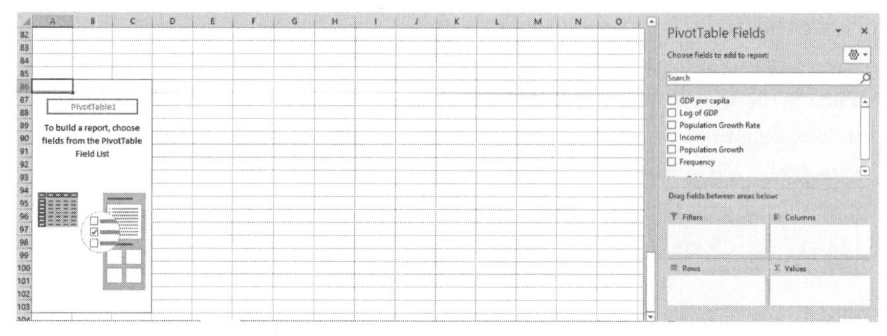

Fig. 8.9 [PivotTable1] appears from **A86**, and [PivotTable Fields] is displayed at the right side of the screen.

iii) Click [OK] (Fig. 8.8).

Then, [PivotTable1] appears from A86, and [PivotTable Fields] is displayed at the right side of the screen (Fig. 8.9). In [PivotTable Fields], six field names, [GDP per Capita], [Log of GDP], [Population Growth Rate], [Income], [Growth Rate], and [Frequency], are displayed. (If some names are not displayed, drag the [Scroll Bar] at the right side of the field names.) Perform the following procedures.

i) Move the mouse pointer to [Income] and press the mouse button, hold the button down, drag [Income] to [Rows] in [Drag fields between areas below:], and release the button. This operation is called "Drag and Drop." The [Income] box is checked and appears in [Rows]. By this procedure, [Income] is selected as a row variable.
ii) In the same way, drag and drop [Population Growth] to [Columns].
iii) Drag and drop [Frequency] to [Σ Values] (Fig. 8.10).

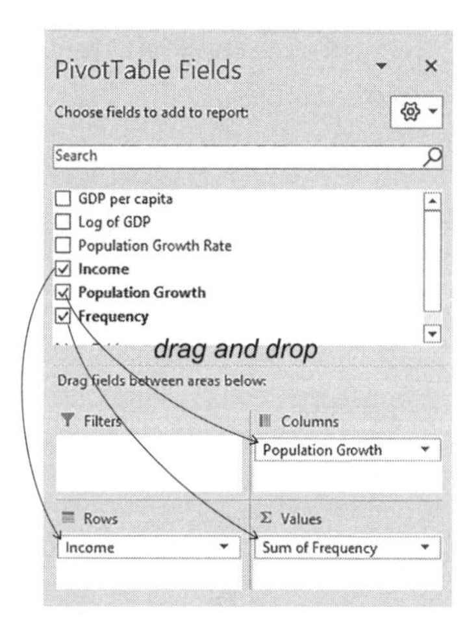

Fig. 8.10 Move the mouse pointer to [Income] and press the mouse button, hold it down, drag [Income] to [Rows] in [Drag fields between areas below:], and release the button. This operation is called "Drag and Drop." The [Income] box is checked and appears in [Rows]. By this procedure, [Income] is selected as a row variable. In the same way, drag and drop [Population Growth] to [Columns]. Drag and drop [Frequency] to [Σ Values].

The table appears in the range from A86. Click an empty cell outside the table. [PivotTable Fields] disappears and the table becomes inactive. If we want to edit the PivotTable, click the table. [PivotTable Fields] appears, and you can edit it.

Input the title of the table, **Table 1 Contingency Table of Income and Population Growth Rate**, in A85. Next, we calculate the ratios to the row totals and make the table in the range from A93. Input **Table 2 Ratios to Row Totals** in A93 and the categories of the variables. For the calculation of the ratios, input **=B88/$E88** combining the relative and absolute cell addresses in B96 and copy it to the range of the table. We can see that the population growth rates tend to be higher in low-income countries (Fig. 8.11).

	A	B	C	D	E
85	Table 1 Contingency Table of Income and Population Growth Rate				
86	**Sum of Frequency**	**Column Labels**			
87	**Row Labels**	**High**	**Low**	**Middle**	**Grand Total**
88	High	1	21	4	26
89	Low	20	23	9	52
90	**Grand Total**	**21**	**44**	**13**	**78**
91					
92					
93	Table 2 Ratios to Row Totals				
94		Population Growth Rate			
95	Income	High	Low	Middle	Grand Total
96	High	3.85%	80.77%	15.38%	100.00%
97	Low	38.46%	44.23%	17.31%	100.00%
98	Grand Total	26.92%	56.41%	16.67%	100.00%

Fig. 8.11 Contingency table and ratios to row totals.

8.3 Summarizing the Data with the PivotTable

Pivot Table is a very convenient tool, and we can use it to summarize the data easily. Here, we obtain the averages of the [Population Growth Rate] for low- and high-income countries. Move the active cell in the List. Click [Insert] table→ [PivotTable]. Select [Existing Work Sheet], input **A101** in [Location], and click [OK]. In [PivotTable Fields] on the right side of the screen, drag and drop [Income] to [Rows] and [Population Growth Rate] to [Σ Values]. Currently, the sums are calculated, so we have to change them to averages. Click the down arrow for [Sum of Popul….] (Fig. 8.12) and select [Values Field Settings] in the submenu (Fig. 8.13). The [Value Field Setting] box appears. Click [Average] in [Choose the type of calculation you want to summarize data from selected field]→ [OK] (Fig. 8.14). The averages of the [Population Growth Rate] in high- and low-income countries are calculated (Fig. 8.15). If we select the wrong field name, click the checked box in front of the field name. The check disappears and the field name is removed from [Drag fields between areas below:].

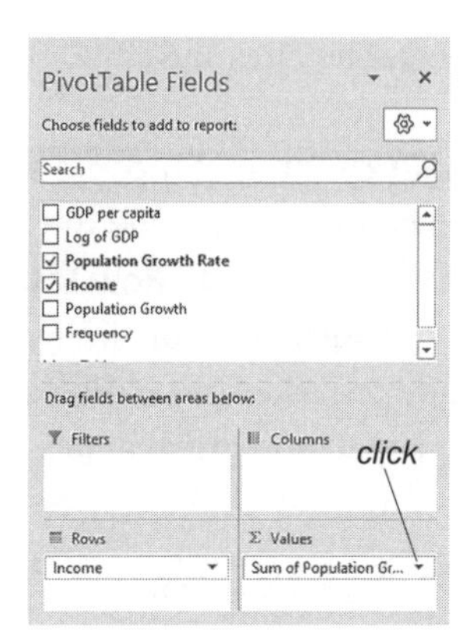

Fig. 8.12 In [PivotTable Fields] on the right side of the screen, drag and drop [Income] to [Rows] and [Population Growth Rate] to [Σ Values]. Currently, the sums are calculated, so we change them to the averages. Click the down arrow for [Sum of Popul….] and select [Values Field Settings] in the submenu.

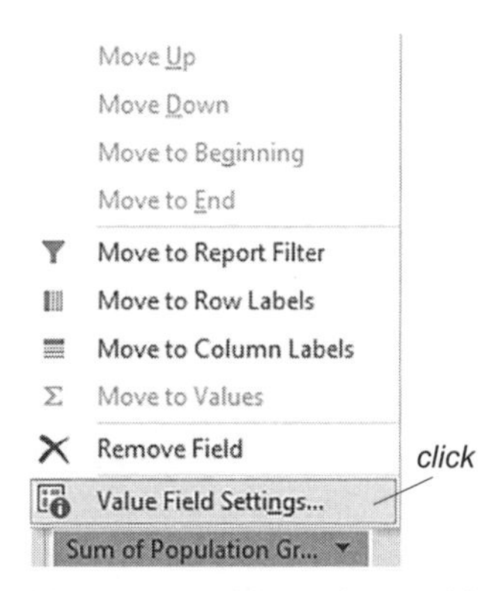

Fig. 8.13 Select [Values Field Settings] in the submenu.

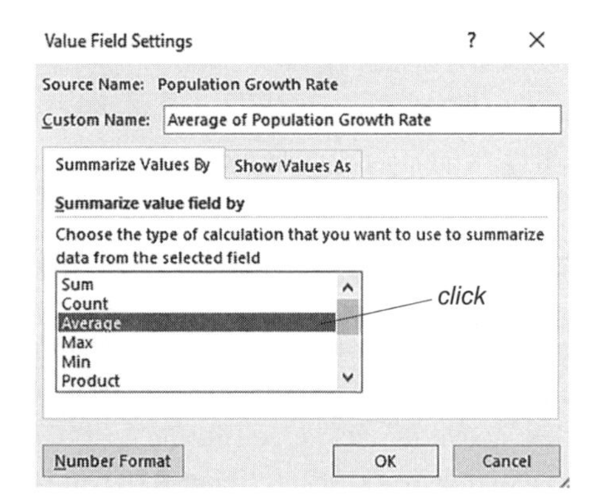

Fig. 8.14 Click [Average] in [Choose the type of calculation you want to summarize data from selected field] → [OK].

	A	B
101	Row Labels	Average of Population Growth Rate
102	High	0.18%
103	Low	0.83%
104	Grand Total	0.61%

Fig. 8.15 The averages of [Population Growth Rate] in high- and low-income countries are calculated.

Previously, we made a [Frequency] column, but we do not need this column now. Move the active cell in the List, and click [Insert] tab→ [PivotTable]. Click [Existing Work Sheet] and input **D101**. Drag and drop [Income] to [Rows], [Growth Rate] to [Columns], and [Population Growth Rate] to [Σ Values]. Click the down arrow for [Sum of Population Gr…]. Select [Values Field Settings] in the submenu. Click [Count] in [Choose the type of calculation you want to summarize data from selected field]→ [OK]. We obtain the same contingency table.

8.4 Calculation of Correlation Coefficient

8.4.1 Correlation Coefficient

The relation of two variables is called correlation. In statistics, when there is a linear correlation between two quantitative variables, we say that the variables are correlated. When the scatter diagrams are as shown in (a) and (b) of Fig. 8.16, and

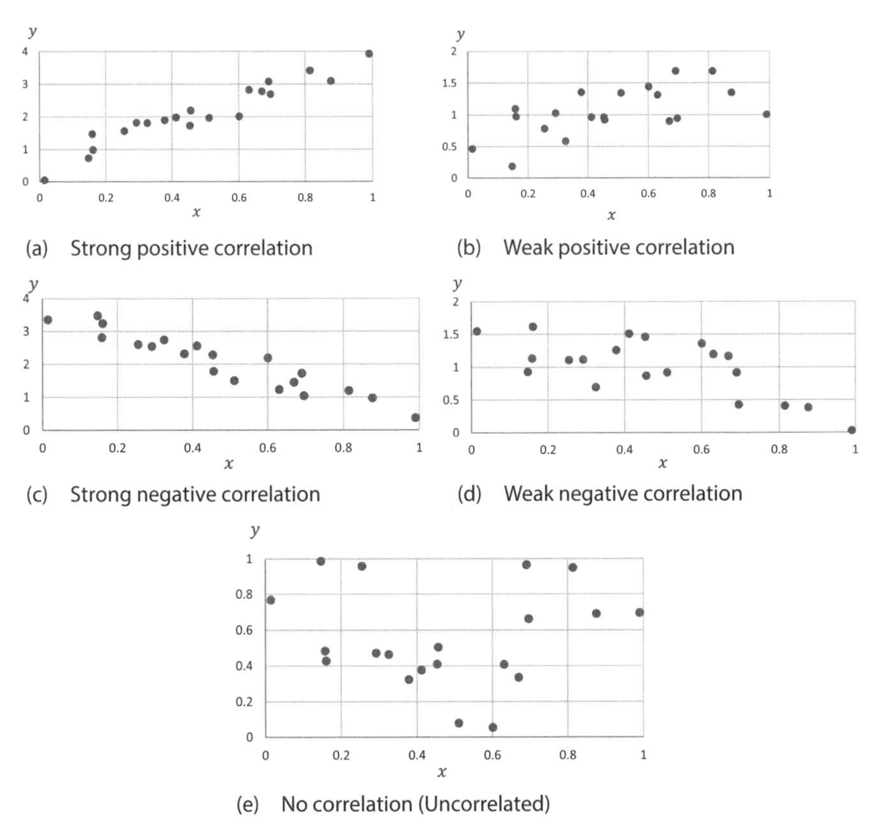

Fig. 8.16 Correlations among variables.

one variable increases as the other variable increases, we call this a positive correlation. On the other hand, when one variable decreases as the other variable increases, we call this a negative correlation as in Fig 8.16 (c) and (d). When there is no relation as in Fig. 8.16 (e), the variables are uncorrelated. In Fig. 8.16 (a) and (c), there are clear linear relations; however, the relations become unclear in (b) and (d). These degrees of correlation are described as "strong" and "weak." In (a) to (d) of Fig. 8.16, there are strong positive, weak positive, strong negative, and weak negative correlations.

The correlation coefficient represents the linear relation between two variables. Here, we consider the product-moment correlation coefficient, which is the most widely used one. The term "correlation coefficient" generally refers to the product-moment correlation coefficient, and hereafter, we will just call it the correlation coefficient. Other correlation coefficients include rank correlation coefficients.

The covariance is a value representing the correlation of two variables that corresponds to the variance of one of the variables. The covariance of x and y, S_{xy}, is obtained from the sum of the products of the deviations of x and y divided by the number of observations, n:

$$S_{xy} = \frac{\{(x_1 - \bar{x})(y_1 - \bar{y}) + (x_2 - \bar{x})(y_2 - \bar{y}) + \cdots + (x_n - \bar{x})(y_n - \bar{y})\}}{n} = \left\{\sum (x_i - \bar{x})(y_i - \bar{y})\right\} \Big/ n \tag{8.1}$$

Here, we divide the sum of the products by n. If $x = y$, we get the variance. In Excel, we can calculate the covariance using the **COVAR** function: **COVAR(range of x, range of y)**. If we exchange x and y, we get the same result.

Since the covariance takes $(-\infty, \infty)$, it is difficult to evaluate the correlation using the covariance. So, we standardize the covariance by dividing by the standard deviations of x and y, S_x and S_y, and we get the correlation coefficient, given by

$$r_{xy} = \frac{S_{xy}}{S_x S_y} = \frac{\sum (x_i - \bar{x})(y_i - \bar{y})}{\sqrt{\sum (x_i - \bar{x})^2}\sqrt{\sum (y_i - \bar{y})^2}} \tag{8.2}$$

The correlation coefficient is always between −1 and 1, satisfies $-1 \le r_{xy} \le 1$, and

i) $r_{xy} = 1$: all observations are on one straight line and the slope of the line is positive (perfect positive correlation)

ii) $r_{xy} > 0$: positive correlation. As r_{xy} approaches 1, the relation becomes stronger.

iii) $r_{xy} < 0$: negative correlation. As r_{xy} approaches −1, the relation becomes stronger.

iv) $r_{xy} = -1$: all observations are on one straight line and the slope of the line is negative (perfect negative correlation).

The correlation coefficient represents the linear relation of two variables. Even if there is a close relation between the variables, the correlation coefficient may be close to zero. For example, suppose that (x, y) are $(-3, 9)$, $(-2, 4)$, $(-1, 1)$, $(0, 0)$, $(1, 1)$, $(2, 4)$, and $(3, 9)$. The data satisfy $y = x^2$ and a close relation is considered to exist between the two variables. However, the correlation coefficient is 0. In statistics, we say that the variables are independent if there is no relation between them. If the variables are independent, the correlation coefficient is zero. However, we cannot say that the variables are necessarily independent even if the correlation coefficient is 0. In a case such that y decreases as x increases initially, but y increases later on, that is, the relation has a U shape, we cannot find the relation with the correlation coefficient. (For analyzing this kind of data, we need to use a scatter diagram or another method.) Moreover, the correlation coefficient expresses whether a linear relation between the variables exists or not, and it does not address causality between the variables (i.e., that one is the cause and the other is the result) .

We can calculate the correlation coefficient easily using the Excel function: **CORREL(range of x, range of y)**. We can get the same result if we change the order of the ranges of x and y.

8.4.2 Calculation of Covariance and Correlation Coefficient Using the Population Data

Let's calculate the covariance and correlation coefficient of [Log of GDP] and [Population Growth Rate]. Excel has functions for calculating these values. However, to fully understand what they mean, we will calculate them from the deviations. Copy the [Log of GDP] and [Population Growth Rate] data including the field names to the range beginning with A111. Since the [Log of GDP] is calculated using equations, use [Paste Values].

Calculate the averages of these two functions in A190 and B190 using the **AVERAGE** function. Then we calculate the deviations. Input **Deviation of x** in C111, **Deviation of y** In D111 and **Product of Deviations** in E111. Then input **=A112–A$190** and copy it to the data range of Columns C and D, C112:D189 (Since we combine the relative and absolute cell addresses, we do not have to input the equation twice.) Input **=C112*D112** in E112, copy the data range in Column E, and obtain the products of the deviations (Fig. 8.17). Using [AutoSum], calculate the sum of the products in E190.

Let's calculate the covariance and correlation coefficient. Input **Standard**

	A	B	C	D	E
111	Log of GDP	Population Growth Rate	Deviation of x	Deviation of y	Product of Deviations
112	3.61	0.71%	-0.25278	0.000981	-0.000248
113	4.16	0.49%	0.29754	-0.001169	-0.000348
114	4.73	0.77%	0.86997	0.001639	0.001426
115	4.68	-0.06%	0.81484	-0.006676	-0.005440
116	3.18	0.32%	-0.67978	-0.002889	0.001964
117	4.64	0.18%	0.77739	-0.004338	-0.003372

Fig. 8.17 Obtain the products of the deviations.

	A	B
195	Standard Deviation of x	0.670527
196	Standard Deviation of y	0.008793
197	Covariance	-0.003974
198	Correlation Coefficient	-0.674067
199		
200	From Excel Functions	
201	Covariance	-0.003974
202	Correlation Coeffcient	-0.674067

Fig. 8.18 Covariance and correlation coefficient.

Deviation of x, **Standard Deviation of y**, **Covariance**, and **Correlation Coefficient** in A195:A198. Input **=STDEVP(A112:A189)** in B195 and **=STDEVP(B112:B189)** in B196 and calculate the standard deviations of two variables. Input **=E190/78** in B197 and calculate the covariance, and input **=B197/(B195*B196)** in E198 to calculate the correlation coefficient.

Finally, we calculate the covariance and correlation coefficient using the Excel functions. Both functions follow the format function name(range of *x*, range of *y*). Input **=COVAR(A112:A189, B112:B189)** in B201 and **=CORREL(A112:A189, B112:B189)** in B202. Label them so they can be understood easily.

8.5 Exercises Using the Countries' Population Data

1. Make a scatter diagram of the log ([Population Density]) and [Population Growth Rate] and consider the relation of the variables.
2. Make a contingency table of the [Population Density] and [Population Growth Rate].
3. Calculate the variance and correlation coefficient of the log([Population Density]) and [Population Growth Rate] using the deviations and Excel functions.

9

Macros and User-Defined Functions

A macro is a set of commands used in Excel. When a procedure consisting of commands needs to be repeated often, we can do so easily if we record the commands as a macro. Then we do not have to retype the commands every time we wish to perform the procedure. Since a macro is a computer program that works on Excel, learning to construct a macro can help you understand general computer programming.

There are two ways to create a macro. One is to record the commands we input by typing or using the mouse. The other is to make a macro using the computer language Visual Basic for Application (VBA). We can make a complicated macro combining both.

Excel has many functions, but not all necessary functions. In some cases, we need to create functions (user-defined functions). User-defined functions are made using VBA.

In the next chapter, we generate random numbers following various distributions and learn about important probability theorems such as the Law of Large Numbers and the Central Limit Theorem in a simulation based on macros and user-defined functions.

VBA is a modification of Visual Basic that is widely used in Windows for applications. What you learn in this chapter can be used in various fields. For details on the use of VBA in Excel, see textbooks in (**Excel, Macros and VBA**) of References.

9.1 Making Macro and Running It

To use macros in Excel, we have to insert the [Developer] tab. Click [File] → [Option] (Fig. 9.1). The [Excel Options] box appears. We select the [Customize Ribbon] tab. Click the box in front of [Developer] in [Main Tabs] so that it is checked (Fig. 9.2). Click [OK] and the [Developer] tab appears in the ribbon.

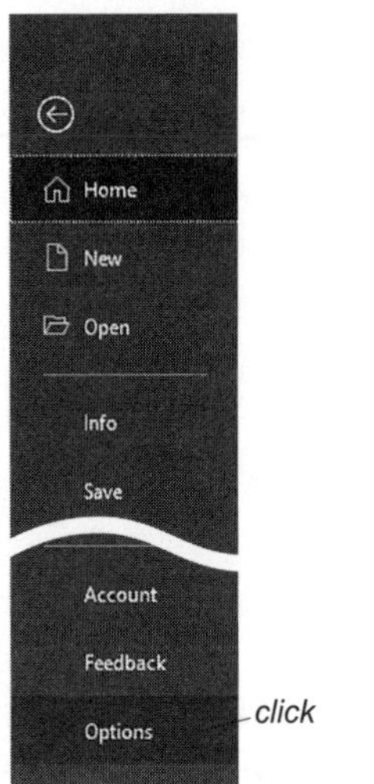

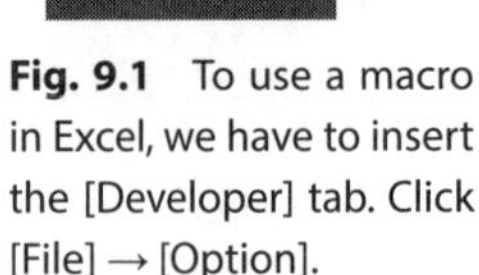

Fig. 9.1 To use a macro in Excel, we have to insert the [Developer] tab. Click [File] → [Option].

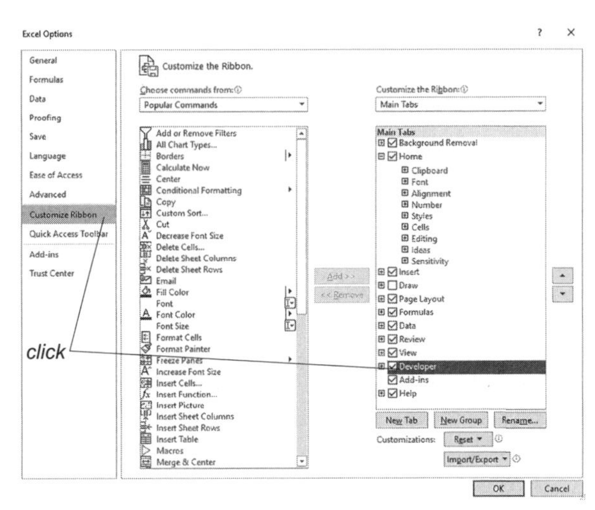

Fig. 9.2 The [Customize Ribbon] tab appears, so click the box in front of [Developer] in [Main Tabs] so that it is checked. Click [OK] and [Developer] appears in the tabs.

9.1.1 Creating a Simple Macro

As an example of a macro, let's create a macro that changes a number to a percentage with 2 digits after the decimal point, and move the active cell down by one cell. We use a new book (if you are using Excel, Click [File] →[New] →[Blank workbook]). Input **0.123** in A1. We make it 12.30%. As described earlier, we click [Percent Style] → [Increase Decimal] twice and press the [Enter] key. Do not start these procedures yet.

We record the procedures. Move the active cell to A1. Click [Developer] tab then the ribbon changes to that of [Developer] (Fig. 9.3). First, click [Use Relative References]. Note that if [Use Relative References] is not selected, the absolute reference is used, and if you do a procedure that moves the active cell from A1 to A2, the command becomes "move the active cell to A2" independent of the position of the active cell. Click [Record Macro] (Fig. 9.4). The [Record Macro] box opens. Input **Percent1** in [Macro name:] and [Shortcut key:] in **a**. In [Shortcut key], the upper- and lower-case alphabets are distinguished. Use the lower case alphabet. Click [OK], then the recording starts (Fig. 9.5). Click the [Home] tab → [Percent Style] → [Increase Decimal] twice and press the [Enter] key. Since the recording is finished, click the [Developer] tab → [Stop Recording] in the Code group (Fig 9.6). The recording will continue until you stop it.

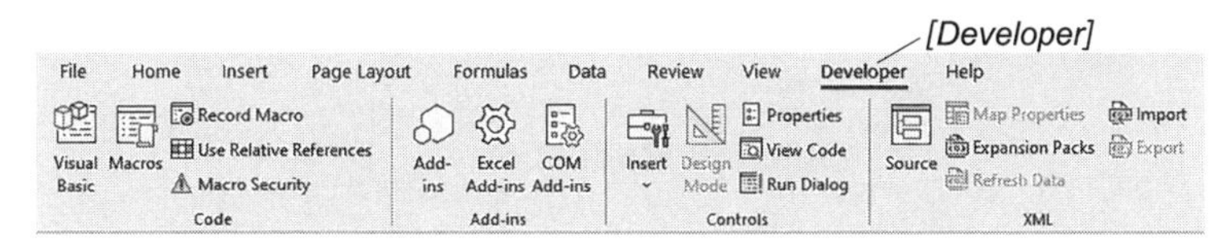

Fig. 9.3 Click the [Developer] tab and the ribbon changes to that of [Developer].

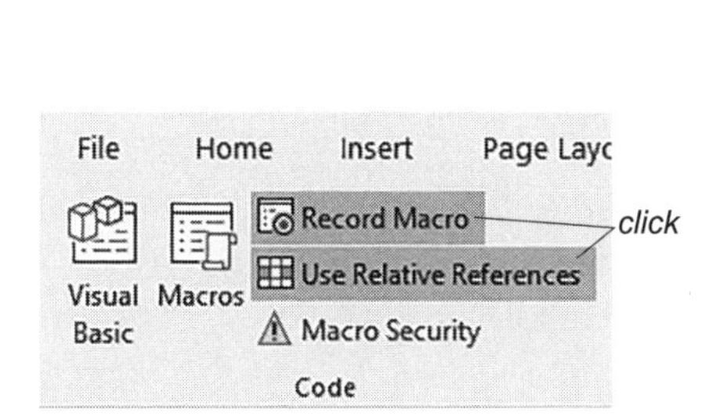

Fig. 9.4 Click [Use Relative References] → [Record Macro].

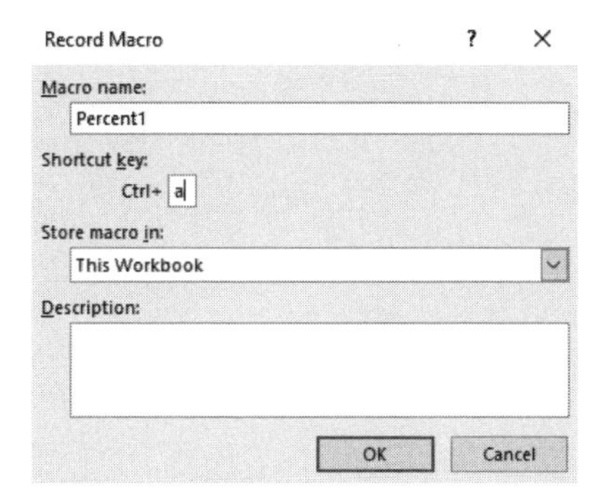

Fig. 9.5 [Record Macro] opens. Input **Percent1** in [Macro name:] and [Shortcut key:] in **a**. Click [OK].

When the macro is recorded, we run it. Input **0.2345** in A2 and move the active cell to A2. Hold the [Ctrl] key down while pressing the [a] key. Hereafter, we will write this action as [Ctrl]+[a]. When the macro is done running, A2 changes to 23.45%, and the active cell moves down by a cell.

Fig. 9.6 To finish recording, click the [Developer] tab → [Stop Recording].

9.1.2 Displaying the Macro

Excel records commands that are input by typing or by using the mouse and converts them to VBA codes when [Record Macro] is started. The VBA codes are written in the [Module] sheet. Click [Macros] in the [Developer] tab (Fig. 9.7). The [Macro] box appears, we select [Percent1] in [Macro name] and click [Edit] (Fig. 9.8). The Visual Basic editor starts and the VBA codes for the commands just input are displayed. Maximize [Modele1 (Code)] by clicking the [Maximize] button at the upper right side of the module (Fig. 9.9).

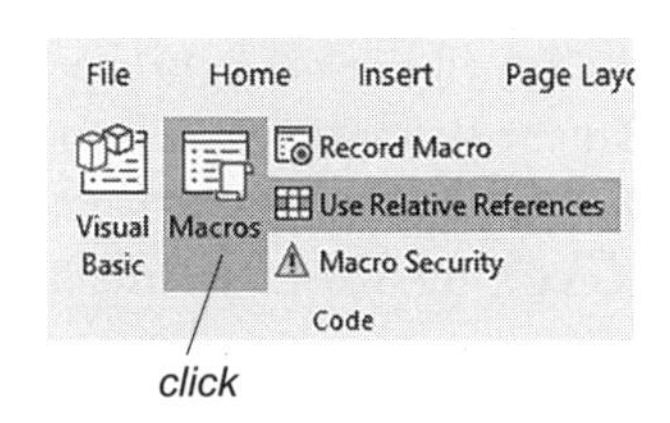

Fig. 9.7 Click [Macros] in the [Developer] tab.

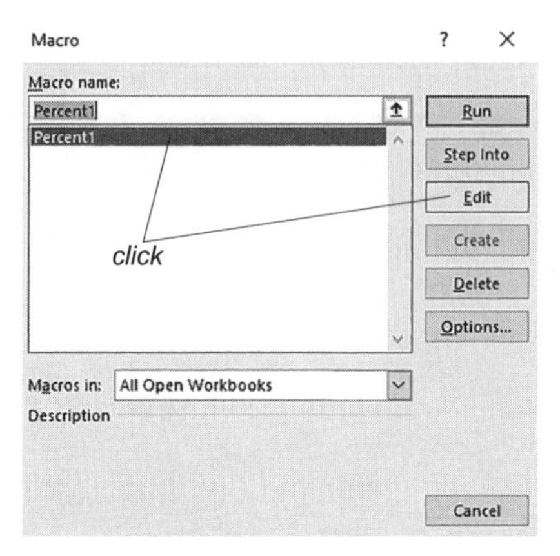

Fig. 9.8 The [Macro] box appears, and we select [Percent1] in [Macro name] and click [Edit].

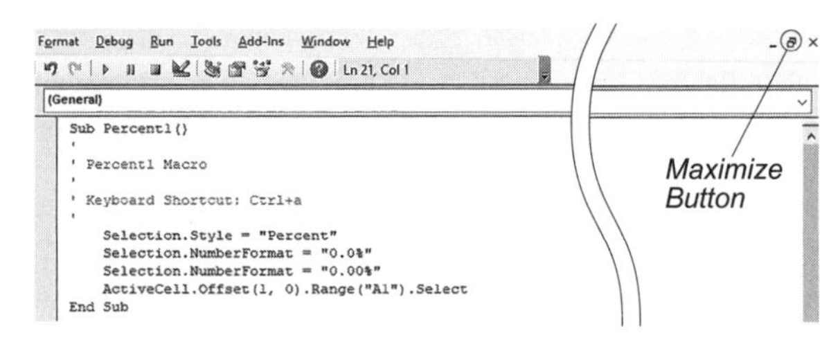

Fig. 9.9 The Visual Basic editor starts and the VBA codes of the commands just input are displayed. Maximize [Modele1 (Code)] by clicking the [Maximize] button at the upper right of the module.

The line stated from [Sub] (the [Sub] statement; a command performed by a computer is called a statement) is the start of the macro, and declare its name. The next several lines beginning with [‘] represent the name of the macro and its short cut key (the key used to execute the macro). The three lines beginning [Selection.] are the statements that change a number into a percentage and increase the number of digits after the decimal point. All of them are in the style of [Selection. xxxxx]. In VBA, we usually specify the object and give the operation for the object after a period.

The statement [ActiveCell.Offset(1, 0).Range("A1").Select] corresponds to the operation that moves the active cell down by one cell. [End sub] in the final line represents the end of the macro.

9.1.3 Changing the Macro Using a Subroutine

When a macro becomes long or the same statements are used repeatedly, we can make the statements into a subroutine. Not only in VBA but also in other computer languages, good usage of subroutines is fundamental to making good computer programs. Make sure that the Visual Basic editor is active and we can edit the macro. Let's make the action part (the four lines from [Selection.Style = "Percent"] to [ActiveCell.Offset(1, 0).Range("A1").Select]) into a subroutine named [Persub] and rewrite the macro so that it works using the subroutine. Add the following statements after [Sub Percent1()] and the five lines starting [']:

Persub
End sub

Sub Persub()

and change the macro as shown in Fig. 9.10. (Upper- and lower-case letters are not distinguished in VBA.)

```
(General)
Sub Percent1()
'
' Percent1 Macro
'
' Keyboard Shortcut: Ctrl+a
'
Persub
End Sub            add these lines
-----------------------------------
Sub Persub()
    Selection.Style = "Percent"
    Selection.NumberFormat = "0.0%"
    Selection.NumberFormat = "0.00%"
    ActiveCell.Offset(1, 0).Range("A1").Select
End Sub
```

Fig. 9.10 When a macro becomes long or the same statements are used repeatedly, we can make the statements into a subroutine. The subroutine is automatically separated by a line.

When we input incorrect statements, we can neither input them in Excel nor correct the macro due to the errors caused by the incorrect statements. In such a case, click [Run] → [Rest] in the Visual Basic editor.

The subroutine is automatically separated by the line. [Persub] calls the subroutine named [Sub Persub()] and runs it, and four statements are performed by the subroutine. [End Sub] is the end of the subroutine.

From the Visual Basic editor, click [File] → [Close and Return to Microsoft Excel], and go back to Excel

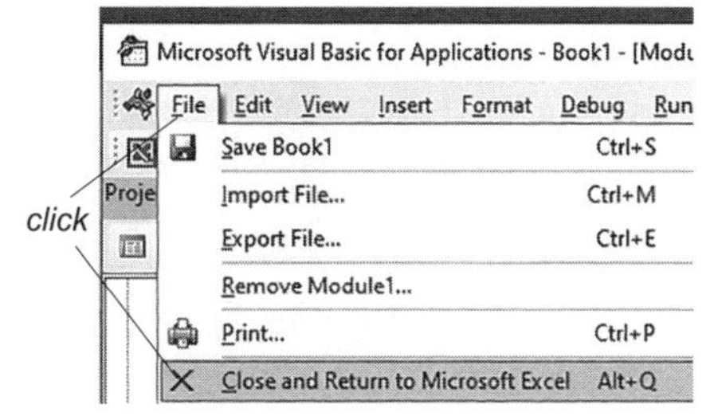

Fig. 9.11 Click [File] → [Close and Return to Microsoft Excel] and go back to Excel.

(Fig. 9.11) and check that the macro works correctly by pressing [Ctrl]+[a]. At this stage, the meaning of the subroutine might be unclear, but it should be made clear by the next example.

In the current macro, we set the number style (percent, two digits after the decimal point) to apply to just one cell. Now, we change it so that it changes the number style of five consecutive cells. Select the [Developer] tab → [Macros] → [Percent1] in the [Macro] box → [Edit] to make it possible to edit the macro. Insert

For i = 1 To 5

at the line before [Persub], and

Next i

at the line after [Persub], so that [Sub Percent1()] becomes

Sub Percent1()
For i = 1 To 5
Persub
Next i
End Sub

The lines **For i = 1 To 5** and **Next i** are rope commands, and they repeat the statements between the two lines five times, increasing *i* by 1 each time. In this case, [Persub] is in the rope, and [Persub] is repeated five times. (Fig. 9.12). Click [File] → [Close and Return to Microsoft Excel] and go back to Excel. Input some numbers in B1:B5, move the active cell to B1, and check whether the macro works by pressing [Ctrl]+[a].

```
(General)
Sub Percent1()
'
' Percent1 Macro
'
' Keyboard Shortcut: Ctrl+a
'
For i = 1 To 5
Persub
Next i
End Sub

Sub Persub()
    Selection.Style = "Percent"
    Selection.NumberFormat = "0.0%"
    Selection.NumberFormat = "0.00%"
    ActiveCell.Offset(1, 0).Range("A1").Select
End Sub
```

Fig. 9.12 Using rope commands, repeat [Persub].

9.1.4 Saving and Opening Files Containing the Macro

The file containing the macro must be saved as [Excel Macro-Enabled Work-

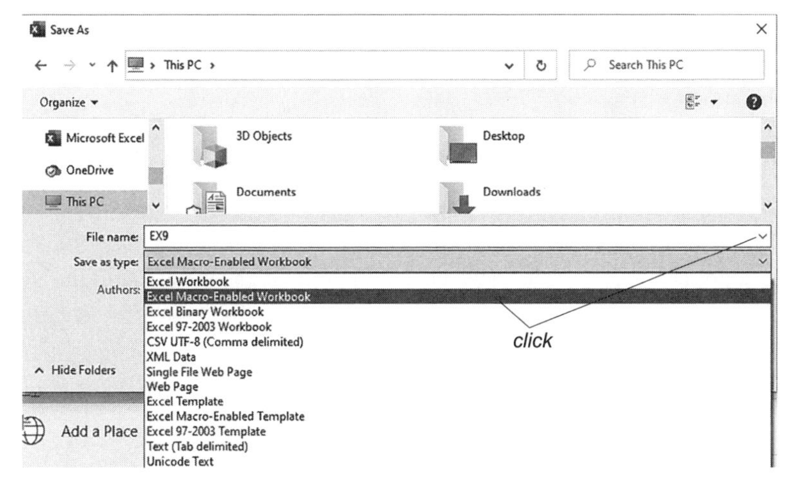

Fig. 9.13 The file containing the macro must be saved as [Excel Macro-Enabled Workbook]. Select [File] → [Save as] → [Save as type:]→ [Excel Macro-Enabled Workbook].

book]. For this purpose, select [File] → [Save as] → [Save as type:]→ [Excel Macro-Enabled Workbook] (Fig. 9.13). Name the file **EX9**, click [Save], and save the file. After saving the file, select [File] → [Close] and close the current workbook. Next, let's open the file containing the macro. Excel presents the message: [SECURITY WARNING Macro has been disabled]. We need to click [Enable Content] to use the macro. As the Excel message warns, the macro may contain computer viruses. So it is important to take care to ensure the security of the macro (Fig. 9.14), especially when you use macros made by other people (especially those you obtain from the internet or as files attached to e-mails).

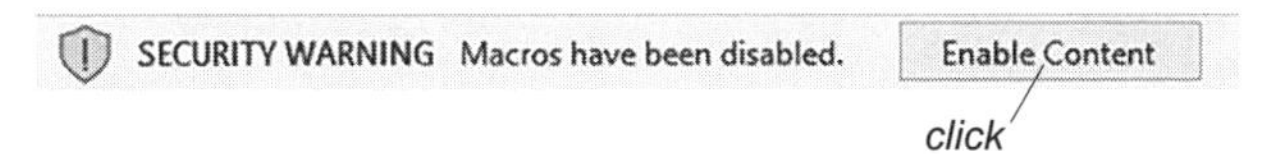

Fig. 9.14 Excel presents the message [SECURITY WARNING Macro has been disabled] and we need to click [Enable Content] to use the macro.

9.2 User-Defined Functions

Excel has many functions but does not include all necessary functions. So, we sometimes need to create a needed function. Such functions are referred to as user-defined functions (custom functions). User-defined functions are written using VBA; however, unlike macros, we have to input them using VBA code.

Here, we first make a function that calculates the factorial of n, and based on the function, we create functions that calculate the numbers of permutations and combinations.

9.2.1 Making a Function that Calculates the Factorial of *n*

Let's create a function that calculates the factorial of n: $n! = n \cdot (n-1) \cdot (n-2) \cdots 3 \cdot 2 \cdot 1$. In a new module, click the [Developer] tab → [VBA] and start the Visual Basic editor (Fig. 9.15). Select [Insert] → [Module] in the menu bar of the Visual Basic editor (Fig. 9.16). Module2 is inserted. Input the following statement:

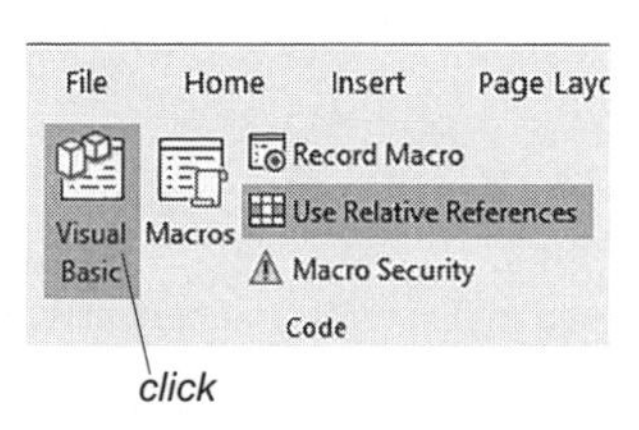

Fig. 9.15 Click [Developer] tab → [VBA] and start the Visual Basic editor.

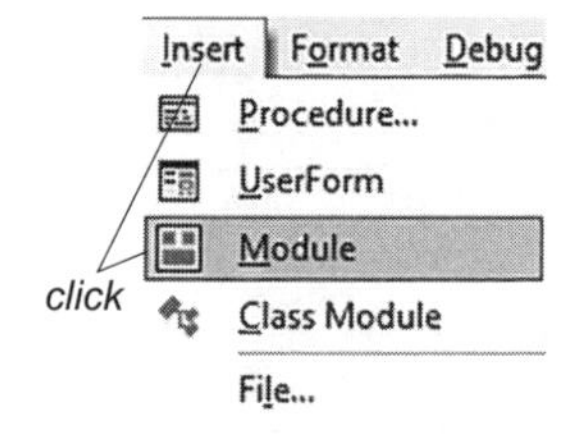

Fig. 9.16 Select [Insert] → [Module] in the menu bar of the Visual Basic editor. Module2 is inserted.

```
Function factorial(n)
  If n = 0 Then
    m = 1
  Else
    m = n
  End If
  k = 1
  For i = 1 To m
    k = k * i
    Next i
  factorial = k
End Function
```

In this function, $n!$ is calculated as follows.

i) In the **Function** statement, we define the function name, and n is an argument of the function.

ii) In the **If** statement, we check whether or not n is zero. We set $m = 1$ if $n = 0$ and $m = n$ otherwise. (0! is defined as 1.) We give the condition in **If**, and the statements after **Then** are executed if the condition is true, but the statements after **Else** are executed if the condition is false. **End If** represents the end of the **If** statement.

iii) Next, we set the initial value of k as 1, and numbers from 1 to m are multiplied one by one by the **For** loop.

iv) The result of the calculation is given as a value of the function by **factorial = k.**

v) **End Function** represents the end of the function.

Go back to Excel (click [File]→ [Close and Return to Microsoft Excel]) and close the Visual Basic editor. We calculate 5!. Input **=factorial(5)** in the proper cell and we obtain $5! = 5\cdot4\cdot3\cdot2\cdot1 = 120$. As the argument value n, we can use a cell address, so that we can obtain $n!$ by changing the number in the cell. The value of $n!$ increases very rapidly as n increases. So, we cannot use this function for large values of n.

9.2.2 A Function that Calculates the Number of Permutations

Suppose that there are n items, and we take out r items considering the order in which they are taken out (permutations). How many permutations are there? The first item can be any of the n items, and there are n different choices. For the second item, there are $n - 1$ different choices since we have already taken out one item, and $n - 2$ different choices for the third item,…., and by the r-th item, there are ${}_nP_r = n\cdot(n-1)\cdot(n-2) \cdots (n-(n-r)) = n!/(n-r)!$ different choices, and ${}_nP_r$ becomes the number of permutations.

Using the function $n!$, we create a function to calculate ${}_nP_r$. Click [Developer] → [Visual Basic] and start the Visual Basic editor. Input the following statements below the **factorial** function in [Module2]. (Put at least one blank line between functions. As subroutines, the functions are automatically separated by lines.)

```
Function perm(n, r)
  nr = n – r
  n1 = factorial(n)
  nr1 = factorial(nr)
  perm = n1 / nr1
End Function
```

Go back to Excel and calculate ${}_5P_2 = 20$ by inputting **=perm(5,2)** in the proper cell. Change the values of n and r to calculate various cases.

9.2.3 Number of Combinations

When calculating the number of permutations, we consider the order in which items are taken out, i.e., (A,B,C), (B,C,A), and (C,A,B) are considered different permutations. But the sets all include the same components, namely A, B and C. So, what is the number of final results (combinations) when we do not consider the order? The number of combinations is given by ${}_nC_r$. The number of permutations is ${}_nP_r$ but there are $r!$ different ways of taking out an item from the same combination of items. Therefore, ${}_nC_r = {}_nP_r/r! = n!/\{(n-r)!r!\}$. ${}_nC_r$ is also called a binomial coefficient.

Start the Visual Basic editor and input the following statements below the **perm** function (put a blank line between functions).

```
Function comb(n, r)
    p = perm(n, r)
    r1 = factorial(r)
    comb = p / r1
End Function
```

In this function, the number of permutations ${}_nP_r$ is divided by $n!$ and ${}_nC_r$ is obtained. Return to Excel, input **=comb(5,2)** in the proper cell, and obtain ${}_5C_2$ =10.

While we created functions that calculate $n!$, the number of permutations, and the number of combinations for practice, Excel has built-in functions to perform these operations, namely **FACT(n)**, **PERMUT(n,r)**, and **COMBIN(n, r)**, respectively.

9.3 Exercises on Making Macros and User-Defined Functions

1. Make a macro that changes the number style to [Scientific] with four digits after the decimal point (change 12345 to 1.2345E+04) and moves the active cell down by one cell.
2. Change the macro made in (1) to perform this procedure for 10 consecutive cells.
3. As we already studied in Chapter 2, the population growth rate per year is given by $r = (P_t/P_0)^{1/t}-1$ where P_0 and P_t are the base population and the population t years later, respectively. Make a user-defined function to calculate r.
4. Make a user-defined function that calculates the sum of the inverse numbers from 1 to n, $1 + 1/2 + 1/3 + \cdots + 1/n = \sum 1/i$.

10

Probability and Random Numbers

A set of all of the elements we are interested in is called a "*population*." (A "*population*" is not a group of people in this context.) Since we use the population data of countries (in this dataset, the word "population" refers to the number of people in a country) in practices and exercises, we will write population in italics when we are referring to the set of all elements. Suppose that there is a country of 100 million persons. The *population* is all of the people in that country. Suppose that we want to know the opinions of the people about a certain policy. Conducting a survey of all elements in the *population* is called a complete survey. However, it is usually very difficult to conduct a complete survey. In such a case, we select a part of the *population* and survey the selected group. The process of selecting the group is called "sampling," the selected group is called the "sample," and the survey conducted is called the sample survey. In public opinion surveys done by media companies, they usually choose several thousand individuals and ask them their opinion. The numbers of elements (in this example, persons) in the *population* and in the sample are called the *population* and sample sizes, and are expressed by N and n, respectively.

However, the sample is just a small part of the *population*. Suppose that the *population* consists of 100 million elements (N = 100 million). We select 1,000 elements (n = 1,000) as the sample. The selected portion n/N is 1/100,000. Even if we conduct a larger survey and select 10,000 elements, n/N is 1/10,000. So, there exists a large uncertainty. To deal with the uncertainty we need mathematical tools for describing "probability."

In this chapter, we learn about probability distributions and perform computer simulations using random numbers that help us to understand the probability distributions. Finally, we learn the Law of Large Numbers and the Central Limit Theorem, which are very important theorems in statistics. We will only learn basic statistical methods and theorems. If readers are interested in the details of these methods, please read other statistical textbooks such as those listed in

(**Statistical Theories and Methodologies**) of References.

10.1 Random Variables and Probability Distributions

10.1.1 Discrete-Type Probability Distributions

Suppose that there is a fair coin and so that heads and tails are equally likely results when we toss the coin; that is, the probability of getting heads or tails is the same, and is 1/2 for each. We assign 1 when heads appears and 0 when tails appears. Let's call the result of the coin toss X. X takes 0 with a probability of 1/2 and 1 with a probability of 1/2. In this example, the variables with their probability are called "Random Variables" and are usually capitalized when written. Next, we toss the coin twice. Let X be the total score. The values that X can take are: 0, 1, and 2, and their probabilities are 1/4, 1/2, and 1/4, respectively. The probability is between 0 and 1 and the sum of all probabilities is always 1.

Generally, the random variable X takes k different values, $\{x_1, x_2,\ldots, x_k\}$; the variable is described as being of the discrete type. k may be infinite but must be scattered and countable like natural numbers $\{1, 2, 3,\ldots\}$.

The probability of $X = x_i$

(10.1) $$P(X = x_i) = f(x_i),\ i = 1, 2, \ldots, k$$

is the probability distribution of X. We can use a subscript, x_i, but we hereafter omit the subscript and just write it as x. The probability of each point is the function of x, so $f(x)$ is called the probability function.

The probability that the random number X is less than or equal to a certain value x, i.e.,

(10.2) $$F(x) = P(X \leq x)$$

is called the cumulative distribution function or just the distribution function. In the case of a discrete type random variable,

(10.3) $$F(x) = \sum_{u \leq x} f(u)$$

where $\sum_{u \leq x}$ represents the sum of all values less than or equal to x. $F(x)$ is defined as all values and it becomes a monotonically increasing step function that jumps at $\{x_1, x_2,\ldots, x_k\}$. It also satisfies $\lim_{x\to-\infty} F(x) \to 0$ and $\lim_{x\to\infty} F(x) \to 1$.

The widely used values representing the probability distributions are the expected value (mean) and the variance. The expected value (mean) represents the location of the distribution and is given by

(10.4) $$E(X) = \sum_{x} x f(x)$$

$\sum_x$ represents the sum of all possible values, and the expected value becomes the weighted average of these values. Following the standard notation in statistics, we refer to the mean as μ.

The variance represents the scatter scale of the distribution and is defined by

(10.5) $$V(X) = \sum_x (x-\mu)^2 f(x)$$

The variance is expressed by σ^2 and the square root of the variance, σ, is called the standard deviation.

Here, we explain the binomial and Poisson distributions.

a. Binomial Distribution

Suppose there is a coin that takes 1 with a probability p and 0 with a probability $q = 1 - p$. We toss the coin and get 1 when heads appears and 0 when tails appears. Let's toss the coin n times (this type of trial is called a Bernoulli trial) and let X be the total score. The values that can be taken are $x = 0, 1, 2, \ldots, n$ and

(10.6) $$f(x) = {}_nC_x p^x q^{n-x} = {}_nC_x p^x (1-p)^{n-x}$$

This distribution is called the binomial distribution and we express it as $Bi(n, p)$ in this book. In the binomial distribution, the expected value is $\mu = p$ and the variance is $\sigma^2 = n{\cdot}p(1-p)$.

b. Poisson Distribution

Suppose that there is a certain amount of radioactive material with a long half-life (for example 0.1 mg of uranium). Let's consider the number of atoms that decay in a fixed period (say, within a second). The probability of decaying is very small for each atom, but there are so many atoms that we will observe some decayed atoms.

Let's consider the case where n is very large, p is very small, and both of them are balanced so that $n{\cdot}p = \lambda$ in the binomial distribution, as in this example. It is difficult to calculate the probability distribution from the binomial distribution. However, the distribution given by

(10.7) $$f(x) = e^{-\lambda}\lambda^x / x!, \ x = 0, 1, 2, 3, \ldots$$

when $n \to \infty, p \to 0$ ($n{\cdot}p = \lambda$ is kept in the limit) by Poisson's Law of Small Numbers. This distribution is called the Poisson distribution. The Poisson distribution is obtained by the limit of the binomial distribution, but it only depends on λ, and we do not have to know n and p individually.

The Poisson distribution is widely used in analyses in various fields in both social and natural sciences. In the Poisson distribution, the expected value is $\mu = \lambda$, the variance is $\sigma^2 = \lambda$, and the expected value is the same as the variance.

10.1.2 Continuous-Type Probability Distributions

When the random variable X is a continuous variable, such as length, time, or area, the potential values that X can take are continuous and infinite. In the continuous case, the density is so high that it becomes uncountable. As a result, if we use the method to assign a probability to each possible point, all probabilities become 0. We cannot define the probability as a discrete variable case. In this case, we consider the probability $P(x < X \leq x + \Delta x)$; that is, the probability X is in the small interval between x and $x + \Delta x$. This probability converges to 0 as Δx becomes smaller. Therefore we divide it by Δx and consider the limit of $\Delta x \to 0$, given by

$$f(x) = \lim_{\Delta x \to 0} P(x < X \leq x + \Delta x)/\Delta x \tag{10.8}$$

This function is called the probability density function (or just the density function). In this book, we do not consider distributions that do not converge in Eq. (10.8). The probability X between a and b (a and b are arbitrary constant numbers satisfying $a < b$) is obtained by the definite integral of $f(x)$ and

$$P(a < X \leq b) = \int_a^b f(x)dx \tag{10.9}$$

The cumulative distribution function, $F(x) = P(X \leq x)$, is obtained by

$$F(x) = \int_{-\infty}^{x} f(u)du \tag{10.10}$$

The expected value μ and variance σ^2 are given by

$$\mu = \int_{-\infty}^{\infty} xf(x)dx \quad \text{and} \quad \sigma^2 = \int_{-\infty}^{\infty} (x-\mu)^2 f(x)dx \tag{10.11}$$

The root of the variance σ is called the standard deviation as before.

Next, we learn the exponential, uniform and normal distributions as examples of continuous-type distributions.

a. Exponential Distribution

Suppose there is a radioactive atom. The exponential distribution expresses the distribution of the time until an atom decays. The probability density and cumulative distribution functions are given by

$$\begin{aligned} f(x) &= a \cdot e^{-a \cdot x} \quad \text{if} \quad x \geq 0, \quad \text{and} \quad 0 \quad \text{if} \quad x < 0 \\ F(x) &= 1 - e^{-a \cdot x} \quad \text{if} \quad x \geq 0, \quad \text{and} \quad 0 \quad \text{if} \quad x < 0 \end{aligned} \tag{10.12}$$

The expected value μ and variance σ^2 are

$$\lambda = \frac{1}{a} \quad \text{and} \quad \sigma^2 = \frac{1}{a^2} = \mu^2 \tag{10.13}$$

The survival probability that an atom does not decay until time x is $1 - F(x) = e^{-ax}$. The time at which the survival probability becomes one-half is called the half-life for radioactive atoms. The half-life satisfies $e^{-ax} = 1/2$ and becomes $\log_e 2/a$.

Note that the Poisson distribution is the distribution of the numbers of incidents in a fixed period when the number of objects is very large but the probability of an incident occurring for each object is very small. The exponential distribution represents the distribution of the times between incidents.

b. Uniform Distribution

The uniform distribution takes each point (more accurately, a small interval around each point) in (a, b) to have the same probability and is given by

$$(10.14)\quad f(x) = \frac{1}{b-a} \quad \text{if} \quad a < x < b, \quad \text{and} \quad 0 \quad \text{if} \quad x \le a \quad \text{or} \quad b \le x$$

The expected value and variance are $\mu = (a + b)/2$ and $\sigma^2 = (b - a)^2/12$.

In this book, we write the uniform distribution as $U(a, b)$. Among uniform distributions, the distribution over $(0, 1)$, $U(0, 1)$ is very important, and we use it when we generate various random numbers.

c. Normal Distribution

The normal distribution is one of the most important distributions used in statistics. This distribution is also called the Gaussian distribution. Many phenomena follow this distribution, and many statistical theories are based on this distribution.

The probability density function is given by

$$(10.15)\qquad f(x) = \frac{1}{\sqrt{2\pi}\sigma} \exp\left\{\frac{-(x-\mu)^2}{2\sigma^2}\right\}$$

The expected value is μ and the variance is σ^2. The function is a mountain-type symmetric shape with respect to μ, and it is expressed as $N(\mu, \sigma^2)$. The distribution of $N(0, 1)$ is especially important and is called the standard normal distribution. The probability density and cumulative distribution function of the standard normal distribution are expressed by $\phi(x)$ and $\Phi(x)$, respectively. We cannot obtain values of $\Phi(x)$ analytically because it is a complicated function. However, a highly accurate formula has been developed and we can obtain the values easily. Excel has a function to calculate $\Phi(x)$.

The normal distribution satisfies the following properties, and it is very easy to handle.

i) When X follows $N(\mu, \sigma^2)$, $aX + b$ follows $N(a\mu + b,\ a^2\sigma^2)$. Therefore, the standardized variable $(X - \mu)/\sigma$ follows the standard normal distribution.

ii) X and Y are independent and follow $N(\mu_x, \sigma_x^2)$, $N(\mu_y, \sigma_y^2)$, and $X + Y$ follows the normal distribution $N(\mu_x + \mu_y, \sigma_x^2 + \sigma_y^2)$.

The importance of the normal distribution will continue to be discussed throughout this book.

10.2 Law of Large Numbers and Central Limit Theorem

The Law of Large Numbers and the Central Limit Theorem are major and very important theorems in statistics and probability. They provide a lot of information about the sum or mean of random variables, regardless of their original distribution. We will learn these theorems in this section.

10.2.1 Law of Large Numbers

Suppose that there is a coin that when flipped has a probability p of showing heads and a probability $q = 1 - p$ of showing tails. We assign 1 if heads appears and 0 if tails appears when we toss the coin. Now, we toss the coin n times (number of trials) and get $X_1, X_2, \ldots, X_n$ where X_i is the result of the trial taking 0 or 1 as its value. All trials are independent. $r = \sum X_i = X_1 + X_2 + \cdots + X_3$ is the number of trials in which we get 1 (success trials), and we get the success rate r/n. Note that the success rate is the average of $X_1, X_2, \ldots, X_n$, $\bar{X} = \sum X_i/n$.

The Law of Large Numbers guarantees that r approaches to the true probability p as n increases. (Mathematically, r/n converges to p in probability; that is, $P(|r/n - p| > \varepsilon) \to 0$ for any value of $\varepsilon > 0$. However, if you are not familiar with probability theorems, just understand that r approaches p.)

The success rate is the average of the trials, and we can generalize the argument to the general random variables. The Law of Large Numbers, a fundamental theorem of statistics, guarantees that the average of independent and identically distributed (i.i.d.) random variables approaches to (converges in probability) the expected value.

This means that if the size of the sample is large enough, we can know the *population* well.

10.2.2 Central Limit Theorem

The Law of Large Numbers shows that the average of i.i.d. variables $\bar{X}$ approaches to (converges in probability) the expected value μ. The Central Limit Theorem shows us how $\bar{X}$ approaches μ. Let $X_1, X_2, \ldots, X_n$ be the i.i.d. random variables with the expected value (mean) 0 and the variance σ^2.

$\bar{X} - \mu$ converges to 0 in probability, but we cannot get information about how it approaches to 0. Therefore, we consider $\sqrt{n}\,(\bar{X} - \mu)$; that is $\bar{X} - \mu$ is multiplied by $\sqrt{n}$. In this case $\sqrt{n}$ increases to infinity and the product does not have to approach to 0. The Central Limit Theorems shows us that the distribution of $\sqrt{n}\,(\bar{X} - \mu)$ approaches the normal distribution independent of the original distribution of X_i

and

(10.16) $$\sqrt{n}(\bar{X} - \mu) \to N(0, \sigma^2) \quad \text{as} \quad n \to \infty$$

In other words, if n is large enough, the distribution of $\sqrt{n}\,(\bar{X} - \mu)$ is approximated by the normal distribution independent of the original distribution. We use the word "asymptotic" when we consider a case in which n is large enough, and the distribution is called the asymptotic distribution. From the properties of the normal distribution, the asymptotic distribution of $\bar{X}$ is obtained by $N\,(\mu, \sigma^2/n)$. Note that the Central Limit Theorem holds for discrete-type random variables. For these variables, the cumulative distribution function of $\bar{X}$ approaches that of the normal distribution.

The Central Limit Theorem shows us that the distributions of the sums or averages of random variables can be approximated by the normal distribution if n is large enough. This is a very important statistical theorem along with the Law of Large Numbers. A problem arises with its accuracy in determining the necessary size of n. This depends on the original distribution. If the original distribution is symmetric with respect to μ, we can get a good approximation for small values of n (for example, 10). However, if the distribution is asymmetric and skewed, we need a relatively large n (for example, 30 or more). If the original distribution is the normal distribution, the distribution of $\bar{X}$ always exactly (not approximately) becomes the normal distribution based on the properties of the normal distribution.

10.3 Making Graphs of the Functions

In this section, we make graphs of the probability, probability density, and cumulative distribution functions of the binomial and normal distributions. To proceed, start Excel.

10.3.1 Probability and Cumulative Distribution Functions of the Binomial Distribution

The probability function of the binomial distribution is

$$f(x) = {}_nC_x p^x q^{n-x} = {}_nC_x p^x (1-p)^{n-x}, \; x = 0, 1, 2, \ldots, n$$

Excel has a function for calculating the binomial coefficient that we will use. Click the [Developer] tab → [Visual Basic]. In the Visual Basic editor, click [Insert] → [Module], insert [Module1], and make a user-defined function for the probability and cumulative distribution functions. Input the user-defined function as in Fig. 10.1.

The first function calculates the probability function, and the second one calculates the cumulative distribution function by adding the probabilities from 0 to x. For the calculation performed in the probability function, we need the binomial coefficient. However, VBA does not have a function to calculate the binomial coefficient, so we use the function in Excel. To use Excel functions in VBA, we put **Application.** in front of the Excel function and make it **Application.Combin**. If we do not insert Application., it will not work. When we finish the input, return to [Sheet1] in Excel.

```
Function bin(n, x, p)
q = 1 - p
a = Application.Combin(n, x)
bin = a * p ^ x * q ^ (n - x)
End Function

Function cbin(n, x, p)
a = 0
For i = 0 To x
b = bin(n, i, p)
a = a + b
Next i
cbin = a
End Function
```

Fig. 10.1 The user-defined functions that calculate the probability and cumulative distribution functions of the binomial distribution.

Let's calculate the probability and cumulative distribution functions of the binomial distribution with $n = 5$ and $p = 0.7$. Input **Binomial distribution** in A1, **n** in A2, **5** in B2, **p** in C2, and **0.7** in D2. Input **Probability function** in B3 and **Cumulative distribution function** in C3. Input **0** to **5** in A4:A9.

Input **=bin($B$2, A4, $D$2)** in B4, and **=cbin($B$2, A4, $D$2)** in C4, copy them to B5:C9, and calculate all values of the functions. Make graphs of the probability and cumulative distribution functions. Then, calculate the expected value and variance and check that $\mu = n{\cdot}p = 3.5$ and $\sigma^2 = n{\cdot}p{\cdot}(1 - p) = 1.05$.

Note that we have made the user-defined functions as practice. Excel has the **BINOMDIST** function for calculating the probability and cumulative density functions. Use **=BINOMDIST(x, n, p, FALSE)** for the probability function and **BINOMDIST(x, n, p, TRUE)** for the cumulative distribution function. Check to ensure that the results of the user-defined and Excel functions are the same.

10.3.2 Probability Density Function and Cumulative Probability Function of Normal Distribution

The probability function of the normal distribution is given by

$$f(x) = \frac{1}{\sqrt{2\pi}\sigma} \exp\left\{\frac{-(x-\mu)^2}{2\sigma^2}\right\}$$

Here, we consider the standard normal distribution, $N(0, 1)$, with $\mu = 0$ and $\sigma^2 = 1$. Excel has a function for calculating the cumulative density function, so we use it, and make graphs. Input **Standard normal distribution** in A21, **x** in A22, **Probability density function** in B22, and **Cumulative distribution function** in C22. Input −3 to 3 by intervals of 0.05 from A23 using the [Fill] feature (input **−3** in

A23 and move the active cell to A23, select [File] in the [Home] tab → [Fill] → [Series], and change [Series in] to [Column], set [Step value] to **0.05**, and set [Stop value] to **3**).

Input **=1/SQRT(2*PI())*EXP(−(A23^2)/2)** in B23 and **=NORMSDIST(A23)** in C23, and copy them to the entire data range. Make scatter diagrams ([Scatter with lines]) using these values.

Under the default setup, Excel recalculates all equations in the worksheet when some operations are done. As a result, the performance of the PC becomes slower. Although it may not be a big problem for newer high-performance PCs, you should change the unnecessary equations to values if this happens.

10.4 Simulations Using Random Numbers

We can run simulations and solve problems using random numbers generated by computers. In this section, we generate random numbers following various distributions, and we learn the Law of Large Numbers and the Central Limit Theorem by looking at graphs made using random numbers.

10.4.1 Random Numbers

The random numbers we can generate in Excel are uniform random numbers that follow the uniform distribution over (0, 1), $U(0, 1)$. On the computer, the random numbers are generated by certain rules, and they are therefore not perfectly random, so they are called pseudorandom numbers. (However, in the scope of this book, we can ignore the difference and treat them as perfect random numbers.) Various random numbers are generated from the uniform random numbers. Here, we generate uniform, binomial, Poisson random numbers.

a. Uniform Random Numbers

Insert a new worksheet and change to [Sheet2]. Input **Uniform random number** in A1. The Excel function generating uniform random numbers is **RAND()**. So, input **=RAND()** in A2. Some number between 0 and 1 appears. Copy it to A201 and generate 200 random numbers. The random numbers change if we perform operations, so change the functions to values. Using the methods we have learned in Chapter 6, make a frequency table and histogram with five classes consisting of 0~0.2, 0.2~0.4, 0.4~0.6, 0.6~0.8, and 0.8~1.0. These frequencies do not vary greatly, as the frequency of each class is around 40.

b. Binomial Random Numbers

Using the uniform random numbers of $U(0, 1)$, we generate binomial random numbers. Let u be a uniform random number of $U(0, 1)$. We assign a value of 1 when $u < p$ and 0 when $u \geq p$, so this value is 1 with a probability of p and 0 with a probability of $1-p$. Therefore, we calculate n different values by this method and obtain their sum, the sum becomes the binomial random numbers following $Bi(n, p)$. (In the simulation, the random numbers are usually represented by lower-case letters.)

Let's make the user-defined function. Click the [Developer] tab → [Visual Basic], start the Visual Basic editor, and input the user-defined function shown in Fig. 10.2.

```
Function binrnd(n, p)
a = 0
For i = 1 To n
a = a + bern(p)
Next i
binrnd = a
End Function
```

```
Function bern(p)
a = Rnd
If a < p Then
bern = 1
Else
bern = 0
End If
End Function
```

Fig. 10.2 User-defined function that generates a binomial random number.

binrnd calls the function **bern** n times and the sum of the obtained values is calculated. **bern** is the function that takes a value of 1 if the value of the uniform random number is smaller than u and 0 otherwise. The function generating the $U(0, 1)$ random number is **Rnd** in VBA. When the input is finished, return to Excel and check whether the function works properly.

We generate 200 binomial random numbers from $Bi(5, 0.5)$. Insert [Sheet3]. Input **5** in A1 and **0.7** in A2. Move the active cell to C1, and click the [Developer] tab → [Use Relative References] → [Record Macro]. Input **brand** in [Macro name] and **a** in [Shortcut key], and click [OK]. Now, we record a macro. Input **=binrnd($A$1, $A$2)** and press the [Enter] key. Next, we change the function to the value. Move the active cell back to C1, click the [Home] tab → [Copy] → downward arrow of [Paste] → [Values (V)], and move the active cell down by one cell. When we finish recording the macro, click [Developer] → [Stop Recording]. Press [Ctrl]+[a] (pressing them simultaneously) and check that the macro

```
Sub brand()
'
' brand Macro
'
' Keyboard Shortcut: Ctrl+a
'
    ActiveCell.FormulaR1C1 = "=binrnd(R1C1, R2C1)"
    ActiveCell.Select
    Selection.Copy
    Selection.PasteSpecial Paste:=xlValues, Operation:=xlNone, SkipBlanks _
        :=False, Transpose:=False
    ActiveCell.Offset(1, 0).Range("A1").Select
End Sub
```

Fig. 10.3 Macro that generates a binomial random number.

works correctly.

Select [Developer] → [Macros] → [brand] → [Edit], and the statements of the macro are displayed (Fig. 10.3). So, we change the macro to calculate 200 binomial random numbers. In front of the line "ActiveCell.FormulaR1C1 = "= binrnd (R1C1, R2C1)"" add

For i= 1 To 200

brand1

Next i

End Sub

Sub brand1()

and change the macro as shown in Fig. 10.4. Pressing the [Ctrl]+[a] keys, generate 200 binomial numbers and make a frequency table and histogram. Compare them to the graph of the probability functions. They are similar shapes. Change n and p, and generate various binomial random numbers.

```
Sub brand()
'
' brand Macro
'
' Keyboard Shortcut: Ctrl+a
'
For i = 1 To 200
    brand1
    Next i
End Sub
```

add these lines

```
Sub brand1()
    ActiveCell.FormulaR1C1 = "=binrnd(R1C1, R2C1)"
    ActiveCell.Select
    Selection.Copy
    Selection.PasteSpecial Paste:=xlValues, Operation:=xlNone, SkipBlanks _
        :=False, Transpose:=False
    ActiveCell.Offset(1, 0).Range("A1").Select
End Sub
```

Fig. 10.4 Macro that generates 200 binomial random numbers.

c. Poisson Random Numbers

Let's generate the Poisson random numbers with $\lambda = 3$ with the macro for generating binomial random numbers and the Law of Small Numbers. Input **1000** in A1 and **0.003** in A2. (Although a larger n is better, when n is too large a lot of computational time is required.) Run the macro and generate 200 Poisson random numbers.

This method can help us understand the Poisson distribution; however, it is not an efficient method. In the exercises, we will generate Poisson random variables by a more efficient method.

d. Normal Random Numbers

The normal distribution is the continuous type, so we generate normal random numbers by the inversion method. Let the cumulative distribution function be $y = F(x)$. We get its inverse function $x = F^{-1}(y)$ with continuous-type variables. Let u be a uniform random variable over (0, 1), $U(0, 1)$. Then the cumulative distribution function of $x = F^{-1}(u)$ becomes $F(c) = P(x \leq c)$. In this case, $F^{-1}(u) \leq c \Leftrightarrow u \leq F(c)$ and

$$P(x \leq c) = P(F^{-1}(u) \leq c) = P(u \leq F(c)) = F(c)$$

Therefore, x represents the random variables following the objective distributions.

In the case of the normal distribution, we cannot obtain the inverse function of the cumulative distribution analytically. However, Excel has a function that calculates it with high accuracy, so we calculate the normal random variables of the standard normal distribution using it. In a proper cell, input **=NORMSINV(RAND())**, copy it to 200 cells, and change them to values. Make a frequency table and histogram and compare them with the graph of the probability density function. Note that the function **NORMSINV** is not perfect when the probability is very small and close to 0 or very large and close to 1, and so errors can occur. So, when the values of the function become smaller than −4.5 or larger than 4.5, make them −4.5 or 4.5. Since these probabilities are very small and less than 1/100000, this seldom happens when generating 200 random numbers.

10.4.2 Simulations of the Law of Large Numbers and the Central Limit Theorem

Here, we study the Law of Large Numbers and the Central Limit Theorem, two basic theorems in statistics, using computer simulations.

a. Law of Large Numbers

We calculate the average of n uniform random numbers over (0, 1), and check that the average approaches the expected value of 0.5. Click [Developer] → [Visual Basic], start the Visual Basic editor, and input the following statements.

```
Function mean1(n)
  a = 0
  For i = 1 To n
    a = a + Rnd
  Next i
  mean1 = a / n
End Function
```

Return to Excel, insert [Sheet4], and check whether this function works.

Next we make a macro that repeats this function 200 times. Input **5** in A1. As before, move the active cell to C1 and click [Developer]. Make sure that [Use Relative References] is selected and click [Record Macro]. Make the [Macro Name] **LLN**, [Shortcut] key **b**, and click [OK]. Input **=MEAN1($A$1)** in C1 and press the [Enter] key. Move the active cell back to C1 and click [Copy] → downward arrow of [Paste] → [Values (V)], and move the active cell down by one cell. Select [Developer] → [Stop Recording]. Check whether this macro works properly by pressing [Ctrl]+[b]. Click [Developer] → [Macro] →[LLN] → [Edit], and add the flowing statement before "ActiveCell.FormulaR1C1 = "=mean1(R1C1)""

For i = 1 To 200

LLNsub

Next i

End Sub

Sub LLNsub()

and change the macro so that the procedure is done 200 times as shown in Fig 10.5.

```
Sub LLN()
'
' LLN Macro
'
' Keyboard Shortcut: Ctrl+b
'
For i = 1 To 200
LLNsub
Next i
End Sub
------------------------------------------------------------
Sub LLNsub()
    ActiveCell.FormulaR1C1 = "=MEAN1(R1C1)"
    ActiveCell.Select
    Selection.Copy
    Selection.PasteSpecial Paste:=xlValues, Operation:=xlNone, SkipBlanks _
        :=False, Transpose:=False
    ActiveCell.Offset(1, 0).Range("A1").Select
End Sub
```


Fig. 10.5 Macro for simulating the Law of Large Numbers.

Return to Excel and calculate the averages of the uniform random numbers of $n = 5$ for 200 times by pressing [Ctrl]+[b]. Calculate the average, standard deviation, minimum, maximum, first quartile (25th percentile) and third quartile (75th percentile) for the 200 calculated random numbers. Make $n = 50$, 500, and 5000, generate 200 random numbers and obtain the average, standard deviation, minimum, maximum, first quartile, and third quartile for the 200 random numbers. We can see that the random numbers obtained approach 0.5, the expected value of $U(0, 1)$, as n increases.

b. Central Limit Theorem

Here, we simulate the distribution of the average approaching the normal distribution using $U(0, 1)$ random numbers. Start the Visual Basic editor and input

```
Function mean2(n)
    a = mean1(n)
    mean2 = (12 * n) ^ 0.5 * (a - 0.5)
End Function
```

In this function, we subtract 0.5 (expected value of $U(0, 1)$) from the average of n random variables and calculate $\sqrt{12 \cdot n}$ (since the variance of $U(0, 1)$ is 1/12) so that the variance becomes 1. Following the procedures we have studied, make a macro ([Macro Name] CLT, [Shortcut key] c) that repeats this function 200 times.

Make $n = 2$, 6, and 12, repeat 200 times and make a frequency table and histogram. We can see the shapes of the graphs approach that of the standard normal distribution as n increases.

To finish, save the workbook as **EX10** using [Excel Macro-Enable Workbook].

10.5 Exercises on Probability Distributions and Simulations Using Random Numbers

1. Graph the probability and cumulative distribution functions of the Poisson distribution with $\lambda = 3$. Excel can calculate these functions. The probability function is **Poisson(x, λ, FALSE)** and the cumulative distribution function is **Poisson(x, λ, TRUE)**.
2. Graph the probability density and cumulative distribution functions of the exponential distribution with $a = 3$.
3. The cumulative distribution function of the exponential distribution is $y = F(x) = 1 - e^{-ax}$, $x \geq 0$. The inverse function becomes $x = F^{-1}(y) = -(1/a) \cdot \log_e (1 - y)$. By the inversion method, generate 200 random numbers following the exponential distribution with $a = 1$ and make a frequency table and histogram. In Excel, the natural log (base e) is given by **LN**(value). When u follows $U(0, 1)$, $1 - u$ also follows $U(0, 1)$. Hence we can directly use u (not necessarily $1 - u$) in this exercise.
4. The method of generating Poisson random numbers in (10.4.1) requires a lot of time and is not an efficient method. When the distribution of the time interval between the occurrence of a certain event (such as radioactive decay) and the occurrence of the next event follows the exponential distri-

bution with $a = 1$, the total number of events that occur in a fixed time t follows the Poisson distribution with $\lambda = t$.

The next function generates the Poisson random numbers using this fact. Using this function, generate 200 Poisson random numbers and make a graph.

```
Function pornd(t)
    x = 0
    t1 = -Log(Rnd)
    Do While t1 < t
        x = x + 1
        t1 = t1 - Log(Rnd)
    Loop
    pornd = x
End Function
```

Unlike in Excel, the **Log** function in VBA is the natural logarithm whose base is e. The **Do while** statement executes the statements to **Loop** while the given condition ($t_1 < t$) is satisfied.

5. Using binomial random numbers of the binomial distribution $Bi(2, 0.7)$, calculate the average of n random numbers and simulate the Law of Large Numbers.
6. Using the binomial random numbers $Bi(2, 0.5)$ and $Bi(2, 0.8)$, simulate the Central Limit Theorem. By approximating by the normal distribution, check that the latter requires a larger value of n.

11

Estimation and Testing of the Normal Population

As we have studied in the previous chapter, a set of all elements we are interested in is called a *population*. (It is not a number of people.) Since we use the Countries' Population Data (in the dataset, the word "population" refers to the number of people in a country) in practices and exercises in this text, we use *population* (in *italics*) when we refer to the set of all elements in this and later chapters. Suppose that the *population* follows a certain distribution given by $f(x)$. Our purpose is to learn about the *population*. Usually, it is very difficult to do a complete survey (surveying all elements of the *population*) since the *population* contains many elements. In that case, we survey by sampling; that is, select a sample from the *population* and conduct a sample survey. We select the sample by simple random sampling in which elements in the *population* have equal probabilities of being chosen; this is the most basic and widely applicable sampling method. Suppose that we select $X_1, X_2, \ldots, X_n$ as the sample. Since the *population* usually consists of many elements, we consider the size of the *population* N to be infinity ($N = \infty$) cases for mathematical simplicity. In these cases, $X_1, X_2, \ldots, X_n$ become independent and identically distributed (i.i.d.) random variables, and their probability distribution is given by $f(x)$.

In this chapter, we consider the normal *population* that follows the normal distribution $N(\mu, \sigma^2)$. μ and σ^2 are the parameters that determine the *population* distributions and are called *population* parameters. μ and σ^2 are respectively the *population* mean and variance representing the location and scatter scale of the distribution of the *population*.

It is known that many datasets follow the normal distribution, and we can easily treat them mathematically. The normal distribution is the basis of many statistical theorems. Furthermore, if the *population* distribution is not normal, we can asymptotically (i.e., approximately when n is large enough) use the methods described in this chapter.

If the *population* parameters are unknown, we have to obtain them from the selected sample $X_1, X_2, \ldots, X_n$.

That process is called "estimation," and the items obtained from the sample are called "estimators." Since the estimators are functions of $X_1, X_2, \ldots, X_n$ and random variables, their values change depending on the sample. In this chapter, we learn about the processes of "estimation and testing" for a normal *population*.

There are some gaps between the contents of this chapter and descriptive statistic methods explained in Chapters 5-8. It can be difficult to understand the methods described in this chapter only from mathematical equations. The practices in which an actual dataset is used should clarify these methods. If some parts of this chapter are difficult to understand theoretically, do the practices of estimation and testing. Doing the practices will help us to understand these topics. In this book, we do not study the theoretical details, so refer to other textbooks such as those listed in (**Statistical Theories and Methodologies**) of References for theoretical details. We treat the Countries' Population Data as sample data that satisfy the required assumptions.

11.1 Point Estimation and Interval Estimation

Estimating a *population* parameter using one estimator or value is called point estimation. We can also estimate the interval that the *population* parameter is in with a predetermined probability. This is called interval estimation. Here, we first consider the point estimation of the *population* mean and variance μ and σ^2, respectively, and then consider the interval estimation.

11.1.1 Point Estimation

To estimate the *population* mean μ, the sample mean

$$\bar{X} = (X_1 + X_2 + \cdots + X_n)/n = \sum X_i/n \tag{11.1}$$

is used (hereafter, I use "mean" rather than average since we consider the *population* and sample at the same time). The *population* variance σ^2 is estimated by

$$s^2 = \sum (X_i - \bar{X})^2/(n-1). \tag{11.2}$$

Unlike in Chapter 7, we divide the sum of the squared deviations by $n - 1$ (we use n for *population* and $n - 1$ for sample). As mentioned in Chapter 7, statistics like the sample mean are used to summarize the data. The estimators considered here are special statistics.

Since $X_1, X_2, \ldots, X_n$ are random variables, $\bar{X}$, s^2 become random variables. The expected values are:

$$E(\bar{X}) = \mu, \; E\left(s^2\right) = \sigma^2$$

We obtain the true values of the *population* parameters. The estimators whose expected values are the true parameter values are called the "unbiased" estimators. When these estimators converge to the true parameter values in probability if $n \to \infty$, we call these estimators "consistent" estimators. Unbiasedness and consistency are important properties for estimators.

The variance of $\bar{X}$ is

$$V(\bar{X}) = \frac{1}{n^2} V\left(\sum X_i\right) = \frac{\sigma^2}{n} \tag{11.3}$$

It becomes smaller as n becomes larger.

When we substitute the actual observed values into the estimator, we can calculate the value of the estimator. We call it an "estimate." The estimate is one possible value of the estimator that is realized. (Note that we get the above results if the distribution of the *population* is not normal.)

11.1.2 Interval Estimation

As we have studied, the *population* mean μ is estimated using the sample mean $\bar{X}$. The estimators from the sample have probabilistic errors. Usually, μ and $\bar{X}$ are not the same value. In the case of a normal *population*, the probability that they have the same value is zero. However, $\bar{X}$ should be near μ (with a high probability). If we consider a certain interval centered on $\bar{X}$, the probability that the interval contains μ should not be low.

In the interval estimation, we obtain an interval $[L, U]$ that contains the *population* parameter at the predetermined probability $1 - \alpha$, so that

$$P(L \leq \mu \leq U) \geq 1 - \alpha$$

This interval is called the "confidence interval," L is called the "lower confidence limit," U is called the "upper confidence limit," and $1 - \alpha$ is called the confidence coeffcient. We can obtain the confidence intervals of μ and σ^2.

a. Confidence Interval of the *Population* Mean

Since $X_1, X_2, \ldots, X_n$ are i.i.d. random variables following $N(\mu, \sigma^2)$, from the properties of the normal distribution, we get

$$\bar{X} \sim N(\mu, \sigma^2/n)$$

"~" means that the random variable follows the given distribution. Therefore, $(\sqrt{n}(\bar{X} - \mu))/\sigma \sim N(0, 1)$. However, the value of σ is unknown, so we replace it with the sample standard deviation s.

It is known that $\sqrt{n}(\bar{X} - \mu)/s$ follows the t-distribution with degrees of freedom given by $n - 1$, $t\,(n - 1)$, in this case. The t-distribution is symmetric with respect to 0 like the standard normal distribution, and it is widely used in interval estimation and hypothesis testing. The degrees of freedom become $n-1$ because $\sum(X_i - \bar{X}) = 0$ and we lose one of the degrees of freedom. The t-distribution becomes the standard normal distribution if the degrees of freedom increase to infinity. Excel has functions for calculating the t-distribution; we use them in interval estimation and hypothesis testing.

In the t-distribution with $n-1$ degrees of freedom, the point at which the probability above that point becomes $100{\cdot}\alpha\%$ is called the percent point and is expressed by $t_\alpha(n-1)$. $t = \sqrt{n}\ (\bar{X} - \mu)/s$ follows the t-distribution, and the t-distribution is symmetric with respect to 0.

$$P\left\{\left|\sqrt{n}(\bar{X} - \mu)/s\right| \le t_{\alpha/2}(n-1)\right\} = 1 - \alpha$$

and we get

$$P\left\{\bar{X} - t_{\alpha/2}(n-1) \cdot s/\sqrt{n} \le \mu \le \bar{X} + t_{\alpha/2}(n-1) \cdot s/\sqrt{n}\right\} = 1 - \alpha$$

Therefore, the confidence interval of the *population* mean μ for the confidence coefficient $1-\alpha$ is given by

$$\text{(11.4)} \qquad \left[\bar{X} - t_{\alpha/2}(n-1) \cdot s/\sqrt{n},\ \bar{X} + t_{\alpha/2}(n-1) \cdot s/\sqrt{n}\right]$$

For the same value of the confidence interval, the width of the confidence interval becomes smaller as n increases, but its order is $1/\sqrt{n}$.

b. Interval Estimation of the *Population* Variance

The sample variance s^2 is calculated as the sum of the squared deviations divided by $n-1$. The sum of the squared deviations divided by σ^2, $\sum(X_i - \bar{X})^2/\sigma^2$, follows the χ^2-(chi-square) distribution with the degrees of freedom given by $n-1$, $\chi^2(n-1)$. χ is the Greek letter chi. Let $u_1, u_2, \ldots, u_k$ be i.i.d. random variables following the standard normal distribution. The squared sum of these variables, $u_1^2 + u_2^2 + \cdots + u_k^2$, follows the χ^2 distribution with degrees of freedom given by k and $\chi^2(k)$. Like the t-distribution, the χ^2-distribution is a very important distribution, and many statistical theorems and methods are based on this distribution.

Let $\chi_\alpha^2(n-1)$ be the percent point above which the probability becomes $100{\cdot}\alpha\%$. We get

$$P\left\{\chi^2_{1-\alpha/2}(n-1) \le \sum(X_i - \bar{X})/\sigma^2 \le \chi^2_{\alpha/2}(n-1)\right\} = 1 - \alpha$$

Unlike the t-distribution, the χ^2 distribution is not symmetric, and we need two (lower and upper) percent points. From this equation, we get

$$P\left\{\sum(X_i - \bar{X})^2/\chi^2_{\alpha/2}(n-1) \le \sigma^2 \le \sum(X_i - \bar{X})^2/\chi^2_{1-\alpha/2}(n-1)\right\} = 1 - \alpha$$

The confidence interval of σ^2 with the confidence coefficient $1-\alpha$ is

$$\text{(11.5)} \quad \left[\sum(X_i - \bar{X})^2/\chi^2_{\alpha/2}(n-1),\ \sum(X_i - \bar{X})^2/\chi^2_{1-\alpha/2}(n-1)\right]$$

11.1.3 Estimation of the *Population* Mean and Variance Using the Population Data.

Let's estimate the *population* mean and variances using the Countries' Population Data. Open the Countries' Population Data file. We have already used up to [Sheet4], so insert [Sheet5] (click the downward arrow for [Insert] in the [Home] tab → [Insert Sheet]. Copy the data for the [Population Growth Rate] and [Income] (for [Income] use [paste values] from A1; we will use [Income] later). We use the data for the [Population Growth Rate] several times, and we name its data range. Select A2:A79 by dragging → [Formulae] tab → [Define Name] and input **Rate** in the [Name] of the [New Name] box and click [OK]. The data range is defined by [Rate].

a. Estimation of the *Population* Mean

From F1 to F6, input **Mean**, **Standard Deviation**, **Sample Size (n)**, **Degrees of Freedom**, **Confidence Coefficient**, and **Percent Point**. Then, input **=AVERAGE(Rate)**, **=STDEV(Rate)**, **=COUNT(Rate)**, **=G3−1**, **95%**, and **=TINV(1−G5, G4)** from G1 to G6. (The sum of the squared deviations is divided by n in **VARP** and **STDEVP**, and divided by $n-1$ in **VAR** and **STDEV** in Excel.) **TINV** is the function for calculating the percent point of the *t*-distribution. Input **TINV(*α*, degrees of freedom)**. We get $t_{\alpha/2}(k)$ when we input **=TINV(*α*, k)**for the *t*-distribution.

Let's estimate the confidence interval of the *population* mean. From F9 to F12, input **Confidence Interval of Population Mean**, **Width*0.5**, **Lower Limit**, and **Upper Limit**. In G10, we calculate $t_{\alpha/2}(n-1)\cdot s/\sqrt{n}$, half the width of the interval, and input **=G2*G6/SQRT(G3)**. Input **=G1−G10** in G11 and **=G1+G10** in G12 and obtain the lower and upper limits of the confidence interval (Fig. 11.1). (Although Excel has the function **CONFIDENCE** that calculates half of the width of the confidence interval, it is based on the standard normal distribution. It has an error, especially when n is small, so we should not use it.)

	F	G
1	Mean	0.61%
2	Standard Deviation	0.88%
3	Sample Size (n)	78
4	Degrees of Freedom	77
5	Confidence Coeffcient	95%
6	Percent Point	1.9913
7		
8		
9	Confidence Interval of Population Mean	
10	Width*0.5	0.20%
11	Lower Limit	0.41%
12	Upper Limit	0.81%

Fig. 11.1 Estimation of *population* mean.

Excel can express up to 15 significant digits. However, using too many digits is not only meaningless, but also gives an incorrect impression about the accuracy of the estimation. So, in the final report, choose a proper number of significant digits depending on the purposes of the study and the accuracy of the data.

b. Estimation of the *Population* Variance

Next, we estimate the *population* variance. We obtain the percent points of the χ^2 distribution. From F15 to F18, input **Chi-Squared Distribution**, **Lower Percent Point**, **Upper Percent Point**, and **Sum of Squared Deviations**. Input **=CHIINV(1−(1−G5)/2, G4)** in G16, **=CHIINV((1−G5)/2, G4)** in G17 and **=DEVSQ (Rate)** in G18. **CHIINV** is the function for calculating a percent point and **DEVSQ** calculates the sum of the squared deviations.

We estimate the sample variance s^2. Input **Variance** in F21 and **=VAR(Rate)** in G21. Next, we obtain the confidence interval of the 95% confidence coefficient. Input **Confidence Interval of Population Variance**, **Lower Limit**, and **Upper Limit** from F24 to F26. We get the lower and upper limits of the intervals by dividing the sum of the squared deviations by the percent point values. Input **=G18/G17** in G25 and **=G18/G16** in G26 and obtain the confidence interval (Fig. 11.2).

	F	G
15	Chi-Sqaured Distribution	
16	Lower Percent Point	54.623
17	Upper Percent Point	103.158
18	Sum of Squared Deviations	0.00603
19		
20		
21	Variance	7.831E-05
22		
23		
24	Confidence Interval of Population Variance	
25	Lower Limit	5.846E-05
26	Upper Limit	1.104E-04

Fig. 11.2 Estimation of *population* variance.

11.2 Hypothesis Testing

In this section, we first learn the basic concepts involved in hypothesis testing and we then learn how to perform hypothesis testing of the *population* mean and variance in a normal *population*.

11.2.1 What Is Hypothesis Testing?

In hypothesis testing, we compare the observed results from a sample dataset and the expected results and verify the propositions concerning the *population*. Suppose there is a dice. If the dice are correctly made, the probabilities of obtaining the numbers 1 to 6 are equal, with each being 1/6. Table 11.1 shows the frequencies of 60 trials of rolling the dice. They do not match the expected values (in this example 10 each) even if the dice is correctly made.

Table 11.1 Results of 60 trials of rolling the dice.

Number on Dice	1	2	3	4	5	6
Frequency	11	16	11	7	8	7

What is important is whether observed results are within the probabilistic errors or not. In statistics, propositions concerning *population* parameters are called hypotheses. When the observed results are outside of the probabilistic errors and we have to conclude that the hypothesis is not correct, we "reject" the hypothesis. In other words, when the probability of obtaining the observed sample is sufficiently low if the hypothesis is correct, we reject the hypothesis. The criterion for the probability is called the significance level and is expressed by α. We reject the hypothesis when the divergence between the observed sample and the hypothesis is significantly large. (Of cause the results depend on the significance level α. The word "significance" alone is not clear enough. We also have to specify the significance level α.)

In general hypothesis testing, we set various conditions to equal the *population* parameters, and we call these conditions the "null hypothesis." We use H_0 to represent the null hypothesis. The hypothesis that is opposed to the null hypothesis is called the "alternative hypothesis" and is expressed as H_1. The null and alternative hypotheses are mutually exclusive and cannot be valid simultaneously. When the null hypothesis is not rejected, we say that the hypothesis is "accepted." Note that "the hypothesis is accepted" just means that the observed results do not contradict the hypothesis, and it does not mean that the hypothesis is positively proved.

In hypothesis testing, there are two different types of errors:

i) Type I error: We reject the null hypothesis when it is correct.

ii) Type II error: We accept the null hypothesis when it is wrong.

Unfortunately, we cannot reduce the probabilities of both errors simultaneously when the sample size n is fixed. In testing, we fix the probability of type I error to be less than or equal to a certain value α and try to minimize the probability of type II error. As already noted, α is called the significance level. In actual testing, 5% and 1% are often selected as values of α. However, we do not have to set α to one of these values, and it is important to select a proper value depending on the purpose of the study. Next, we will learn about the tests of the *population* mean and variance in a normal *population*.

11.2.2 Testing the *Population* Mean

Tests of the *population* mean μ are the most widely used tests. Here, we learn the two-tailed and one-tailed tests.

a. Two-Tailed Test

In the two-tailed test, the null and alternative hypotheses are given by

$$H_0 : \mu = \mu_0, \ H_1 : \mu \neq \mu_0$$

μ_0 is the value that we obtain depending on the purpose of the test. Suppose that we set the temperature of an air conditioner to 25°C (= 77°F). If the air conditioner works properly, the observed temperature would fall near 25°C (it may vary depending on the weather and the conditions of the room). Therefore, H_0: μ = 25.0.

The test is based on the distance between μ_0 and $\bar{X}$. As we have already studied in the interval estimation, $t = \sqrt{n}(\bar{X} - \mu)/s$ follows a t-distribution with degrees of freedom given by $n - 1$, $t\ (n - 1)$. If the null hypothesis is correct, $\mu = \mu_0$. Therefore, under the null hypothesis (i.e., when the null hypothesis is correct),

$$t = \sqrt{n}(\bar{X} - \mu_0)/s \tag{11.6}$$

follows $t\ (n - 1)$. The statistics used in testing are called test statistics, like t in Eq. (11.6).

Select a proper value of α based on the purpose of the study. In the two-tailed test, we compare t and the percent point of the t-distribution, $t_{\alpha/2}(n - 1)$, and

when $|\ t\ | > t_{\alpha/2}(n - 1)$, we reject the null hypothesis,

when $|\ t\ | \leq t_{\alpha/2}(n - 1)$, we do not reject (accept) the null hypothesis.

In this test, the rejection regions are on both tails (the null hypothesis is rejected if t is too small or too large). Therefore, the test is called the two-tailed test.

We can calculate the probability that the absolute value is larger than $|\ t\ |$ in the t-distribution with degrees of freedom given by $n - 1$, $t\ (n - 1)$. We call it the p-value. We can also perform a test to compare the p-value and α (the null hypothesis is rejected when $p < \alpha$).

b. One-Tailed Test

When we can predict that the value of μ is greater or less than μ_0, we use the one-tailed test. Suppose the value of μ is expected to be larger than μ_0. In this case, the null and alternative hypotheses are given by

$$H_0 : \mu = \mu_0,\ H_1 : \mu > \mu_0$$

We perform the right one-tailed test, in which the rejection region is the right tail of the distribution. (In some books, the null hypothesis is given by H_0: $\mu \leq \mu_0$ but the test is the same.)

Since the null hypothesis is the same,

$$t = \sqrt{n}(\bar{X} - \mu_0)/s$$

follows $t\ (n - 1)$ under the null hypothesis. However, since the alternative hypothesis is different, the rejection region becomes different. Let α be the significance level. In the right one-tailed test, we compare t and $t_\alpha(n - 1)$.

When $t > t_\alpha(n - 1)$, we reject the null hypothesis.

When $t \leq t_\alpha(n - 1)$, we do not reject (we accept) the null hypothesis.

When we expect that μ is smaller than μ_0, we use the hypotheses H_0: $\mu = \mu_0$, H_1:

$\mu < \mu_0$ and perform the left one-tailed test.

When $t < -t_\alpha(n-1)$, we reject the null hypothesis.

When $t \geq -t_\alpha(n-1)$, we accept the null hypothesis.

Let H_1: $\mu > \mu_0$. We can consider the probability that the value becomes larger than the t obtained in $t\,(n-1)$. This probability is called the one-tailed p-value. (When H_1: $\mu < \mu_0$, we consider the probability to be smaller than t.) The p-value represents the smallest significance value at which the null hypothesis is rejected. We can perform the test by comparing the p-value and α (when the p-value $< \alpha$, the null hypothesis is rejected) in the test.

c. Selection of Two-Tailed or One-Tailed Tests

We have studied two-tailed and one-tailed tests. The next question to discuss is how to determine which of them should be used in a given case. Generally, we use the two-tailed test if the *population* mean equals a certain targeted value. For example, if an air conditioner works properly, the temperature of the room should be close to the preset temperature (although it may fluctuate a little bit due to the weather and the conditions of the room). If there is a large difference between the preset and actual temperatures, we conclude that the air conditioner does not work properly. In this case, we use the two-tailed test.

On the other hand, we use a one-tailed test when we can predict the mean value of the *population* theoretically or empirically. Suppose we are taking lectures on statistics, and we take tests before and after the lectures. If the lectures are effective, the test scores after the lectures should be higher than those before the lectures. In this case, we want to know whether the scores are improved (not just changed), and we use the one-tailed tests to determine this.

11.2.3 Tests of the *Population* Variance

$\sum(X_i - \bar{X})^2/\sigma^2$ follows the chi-squared distribution with degrees of freedom given by $n-1$ and $\chi^2(n-1)$. Let's consider the null hypothesis related to σ^2, H_0: $\sigma^2 = \sigma_0^2$. If the null hypothesis is correct,

$$\chi^2 = \sum(X_i - \bar{X})^2/\sigma_0^2 = (n-1)\cdot s^2/\sigma_0^2 \quad (11.7)$$

follows $\chi^2(n-1)$. The test of the *population* variance uses this relationship. Let α be the significance level of the test. We compare the percent points of $\chi^2(n-1)$ and the value of χ^2.

i) When the alternative hypothesis is H_1: $\sigma^2 \neq \sigma_0^2$, we perform the two-tailed test. We accept H_0 if $\chi^2_{1-\alpha/2}(n-1) < \chi^2 < \chi^2_{\alpha/2}(n-1)$ and reject H_0 otherwise.

ii) When H_1: $\sigma^2 > \sigma_0^2$, we perform the right one-tailed test. We reject H_0 if

$\chi^2 > \chi^2_\alpha(n-1)$ and accept H_0 otherwise.

iii) When H_1: $\sigma^2 < \sigma_0^2$, we perform the left one-tailed test. We reject H_0 if $\chi^2 < \chi^2_{1-\alpha}(n-1)$ and accept H_0 otherwise.

Selection of the one-tailed test or two-tailed test depends on whether the size of the *population* variance is predicted either theoretically or empirically. In the variance tests, we can also consider the p-value and perform the tests.

The χ^2 distribution is not symmetric with respect to 0. The two-tailed p-values are obtained by

i) If $F_0(\chi^2) < 1/2,\ 2F_0(\chi^2)$

ii) If $F_0(\chi^2) \geq 1/2,\ 2\{1 - F_0(\chi^2)\}$

where F_0 is the cumulative distribution function of $\chi^2(n-1)$. In the *population* mean case, we can test the hypothesis by comparing the p-value and α. (If the p-value $< \alpha$, we reject the null hypothesis.)

11.2.4 Tests of the *Population* Mean and Variance Using the Population Data

a. Tests of the *Population* Mean μ

In this section, we test the hypotheses concerning the *population* mean μ using the [Population Growth Rate] data. If some item increases by 0.7% per year, it almost doubles in 100 years. We test whether μ is 0.7% or not. The null and alternative hypotheses are

$$H_0 : \mu = 0.7\%, \quad H_1 : \mu \neq 0.7\%$$

Input **Test of Population Mean** in I1, and **Test1 (Two-Tailed Test), Null hypothesis H0, Alternative hypothesis H1, Significance Level, Test Statistic t**, and **Percent Point** in I3:I8. Input **0.7%**, the value of the null hypothesis in J4, **<>0.7%** ("<>" means "≠" in Excel) in J5, and **5%**, the significance level, in J6.

The values of $\bar{X}$, s, n are already calculated. Input **=SQRT(G3)*(G1−J4)/G2** in J7 and calculate the value of the test statistic t. Input **=TINV(J6, G4)** in J8 and calculate the percent point $t_{\alpha/2}(n-1)$. The result is $|t| = 0.890 < t_{\alpha/2}(n-1) = 1.991$, and we accept (do not reject) the null hypothesis (Fig. 11.3).

	I	J
1	Test of Population Mean	
2		
3	Test1 (Two-Tailed Test)	
4	Null hypothesis H0	0.70%
5	Althernative hypothesis H1	<>0.7%
6	Significance Level	5%
7	Test Statistic t	-0.89045
8	Percent Point	1.991254
9	|t|< percent point	
10	Result	Accept H0

Fig. 11.3 Test of the *population* mean (two-tailed test).

The annual world population growth rate between 1960 and 2017, obtained from Table 1.1 in the exercises in Chapter 1, is 1.61%. We test whether the growth rate between 2020 and 2070 becomes lower than this value. In this case, we use the one-tailed test and the null and alternative hypotheses become

$$H_0 : \mu = 1.61\%, \quad H_1 : \mu < 1.61\%$$

We perform the left one-tailed test. In I12:I17, input **Test2 (One-tailed test), Null hypothesis H0, Alternative hypothesis H1, Significance Level, Test Statistic t,** and **Percent point**. Input **1.61%** in J13, **<1.61%** in J14, and **5%** in J15. As before, calculate the test statistic in J16. To obtain the percent point of α, $t_\alpha(n-1)$, we have to use **TINV(2*α, n−1)**, so input **=TINV(2*J15, G4)** in J17. Since $t = -9.972 < t_\alpha(n-1) = -1.665$, we reject the null hypothesis, and it is shown that the population growth rate becomes smaller (Fig. 11.4).

b. Test of *Population* Variance

Using the [Population Growth Rate] data, we test whether or not the *population* variance σ^2 is $(1\%)^2 = 0.0001$. Let

$$H_0 : \sigma^2 = (1\%)^2, \quad H_1 : \sigma^2 \neq (1\%)^2$$

Input **Test of Population Variance** in I22. Input **Test3 (Two-tailed test), Null hypothesis H0, Alternative hypothesis H1, Significance Level, Test Statistic chi2, Lower Percent Point**, and **Upper Percent Point** in I24:I30. From J25 to J27, input **0.0001, <> 0.0001**, and **5%**. The sum of the squared deviations is already calculated, so obtain the test statistic **chi2** dividing the sum of the squared deviations by the value of the null hypothesis. Using the **CHIINV** function, obtain the lower and upper percent points of $\chi^2(n-1)$. $\chi^2_{1-\alpha/2}(n-1) = 54.623 < \chi^2 = 60.302 < \chi^2_{\alpha/2}(n-1) = 103.158$ and the null hypothesis is accepted at the 5% level (Fig. 11.5).

	I	J
12	Test2 (One-tailed test)	
13	Null hypothesis H0	1.61%
14	Alternative hypothesis H1	<1.61%
15	Significance Level	5%
16	Test Statistic t	-9.9721404
17	Percent poit	1.66488454
18	t<-percent poit	
19	Result:	Reject H0

Fig. 11.4 Test of the *population* mean (one-tailed test).

	I	J	K
22	Test3 (Test of Population Variance)		
23			
24	Test3 (Two-tailed test)		
25	Null hypothesis H0	0.0001	
26	Alternative hypothesis H1	<>0.0001	
27	Significance Level	5%	
28	Test Statistic chi2	60.30241	
29	Lower Percent Point	54.62336	
30	Upper Percent Point	103.1581	
31	Lower Percent Point < chi2 < Upper Percent Point		
32	Result	Accept H0	

Fig. 11.5 Test of *population* variance.

11.3 Test of Identity of Two Different Normal *Populations* (Two-Sample Test)

It is very important to determine whether the distributions of two different *populations* are the same or not. For example, when we investigate the effects or side effects of a medicine, we divide the laboratory mice into two groups. We give a medicine to one group and do not give the medicine to the other group and test whether differences (such as the weights of the mice) exist between the two groups. This test is called the Two-Sample Test. In this section, we first learn what the tests of the difference between the two *populations* mean. Then we learn how to test the variances based on the F-distribution and how to test the differences between the *populations*.

11.3.1 Tests of the Difference between the *Populations*

Suppose that the two *populations* follow the normal distributions $N(\mu_1, \sigma_1^2)$, $N(\mu_2, \sigma_2^2)$, and select $X_1, X_2, \ldots, X_m$ from the first *population* and $Y_1, Y_2, \ldots, Y_n$ from the second *population*. We want to test whether $\mu_1 = \mu_2$ or not, and the null hypothesis is given by

$$H_0 : \mu_1 = \mu_2$$

The alternative hypothesis becomes

$$\text{Two-tailed test}: \; H_1 : \mu_1 \neq \mu_2$$

$$\text{One-tailed test}: \; H_1 : \mu_1 > \mu_2 \quad \text{or} \quad H_1 : \mu_1 < \mu_2$$

As before, two-tailed and one-tailed tests are used based on the purpose and on previous information that has been obtained. The test differs depending on whether the *population* variances are the same or not. So, we need to learn both of them.

a. When $\sigma_1^2 = \sigma_2^2 = \sigma^2$

When the two *population* variances are the same and $\sigma_1^2 = \sigma_2^2 = \sigma^2$, the variance σ^2 is estimated by

$$s^2 = \left\{ \sum_{i=1}^{m} (X_i - \bar{X})^2 + \sum_{j=1}^{n} (Y_i - \bar{Y})^2 \right\} / (m + n - 2) \tag{11.8}$$

where $\bar{X}$, $\bar{Y}$ are the sample means of each sample. Under the null hypothesis,

$$t = (\bar{X} - \bar{Y}) / \left\{ s \cdot \sqrt{(1/m) + (1/n)} \right\} \tag{11.9}$$

follows the t-distribution with degrees of freedom given by $m + n - 2$. Therefore, the test is done as follows.

i) In the two-tailed test, we reject the null hypothesis if $|t| > t_{\alpha/2}(m+n-2)$ and accept it otherwise.

ii) When H_1: $\mu_1 > \mu_2$, we reject the null hypothesis if $t > t_\alpha(m+n-2)$ and accept it otherwise.

iii) When H_1: $\mu_1 < \mu_2$, we reject the null hypothesis if $t < -t_\alpha(m+n-2)$ and accept it otherwise.

b. When $\sigma_1{}^2 \neq \sigma_2{}^2$

When the *population* variances are not equal, let

(11.10) $$t = (\bar{X} - \bar{Y})/\sqrt{s_1^2/m + s_2^2/n}$$

where $s_1{}^2 = \sum(X_i - \bar{X})^2/(m-1)$ and $s_2{}^2 = \sum(Y_j - \bar{Y})^2/(n-1)$ are the sample variances of each sample. Then t approximately follows a t-distribution (unfortunately, we cannot obtain the exact distribution). Let

(11.11) $$\nu = (s_1^2/m + s_2^2/n)^2/\{(s_1^2/m)^2/(m-1) + (s_2^2/n)^2/(n-1)\}$$

and let ν^* be the closest integer to ν. The degrees of freedom are given by ν^*. This means that the test statistic t defined in Eq. (11.10) approximately follows $t(\nu^*)$.

i) In the two-tailed test, we reject the null hypothesis if $|t| > t_{\alpha/2}(\nu^*)$ and accept it otherwise.

ii) When H_1: $\mu_1 > \mu_2$, we reject the null hypothesis if $t > t_\alpha(\nu^*)$ and accept it otherwise.

iii) When H_1: $\mu_1 < \mu_2$, we reject the null hypothesis if $t < -t_\alpha(\nu^*)$ and accept it otherwise.

This test is called Welch's test.

Since this test is only approximately valid, the accuracy of the test becomes lower if we use this test when the variances are the same.

11.3.2 Test of Variances and *F*-Distribution

The tests of the *population* means depend on the variances of the *populations*. Moreover, the variances themselves sometimes become important. For example, when we compare the scatter scales of two production processes, the comparison of variances is very important. The null hypothesis is

$$H_0 : \sigma_1^2 = \sigma_2^2$$

The alternative hypothesis is

$$\text{Two-tailed test} : H_1 : \sigma_1^2 \neq \sigma_2^2$$

$$\text{One-tailed test} : H_1 : \sigma_1^2 > \sigma_2^2 \quad \text{or} \quad H_1 : \sigma_1^2 < \sigma_2^2$$

For the test, we consider the ratio of the sample variances and use the F-distribution.

Suppose that two random variables Z_1 and Z_2 satisfy the following three conditions.

i) $Z_1 \sim \chi^2(k_1)$

ii) $Z_2 \sim \chi^2(k_2)$

iii) Z_1 and Z_2 are independent.

In this case,

$$F = \frac{Z_1/k_1}{Z_2/k_2}$$

follows an F-distribution with degrees of freedom (k_1, k_2). Note that when $t \sim t(k)$, t^2 follows $F(1, k)$.

Under the null hypothesis, it is known that

$$F = s_1^2/s_2^2 \tag{11.12}$$

follows $F(m-1, n-1)$. Therefore, the test is done by comparing the value of F and the percent points of the F-distribution with the degrees of freedom $(m-1, n-1)$; they are $F_{\alpha/2}(m-1, n-1)$, $F_{1-\alpha/2}(m-1, n-1)$, $F_{\alpha}(m-1, n-1)$

i) In the two-tailed test, we reject the null hypothesis if $F < F_{1-\alpha/2}(m-1, n-1)$, $F > F_{\alpha/2}(m-1, n-1)$ and accept it otherwise.

ii) When H_1: $\sigma_1^2 > \sigma_2^2$, we reject the null hypothesis if $F > F_{\alpha}(m-1, n-1)$ and accept it otherwise.

iii) When H_1: $\sigma_1^2 < \sigma_2^2$, we reject the null hypothesis if $F < F_{1-\alpha}(m-1, n-1)$ and accept it otherwise.

We can also calculate the p-values by the same methods used for the χ^2 distribution. In Excel, we can get the $100\cdot\alpha$ percent point of $F(k_1, k_2)$ using **=FINV(*α*, k1, k2)**.

11.3.3 Two-Sample Tests Using the Population Data

Let us divide the data into two groups. One group includes the low-income countries (Group1) and the other includes the high-income countries (Group2). For the test, we need the sample size, sample mean, sample variance, and sum of the squared deviation of each group, and obtaining these values is rather bothersome. Excel has a procedure that calculates these numbers in [Data Analysis].

First, we divide the data following the methods studied in Chapter 5. Choose the data for the low- and high-income countries separately. Copy the field name [Income] to A85 and input **Low** in A86.

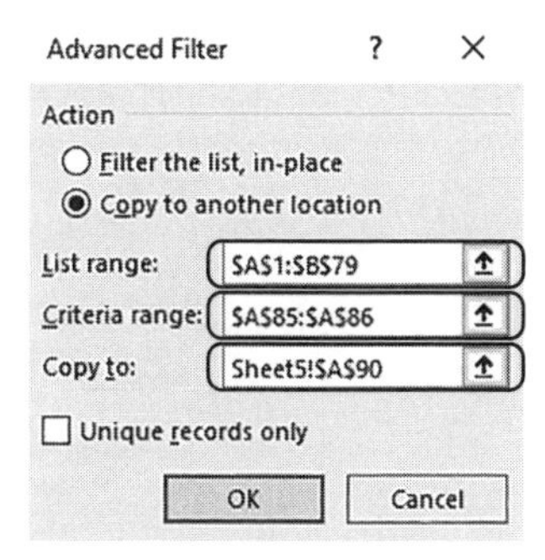

Fig. 11.6 Input conditions in [Advanced Filter].

Move the active cell in the list and click the

	A	B	C	D
90	Population Growth Rate	Income	Population Growth Rate	Income
91	0.71%	Low	0.77%	High
92	0.49%	Low	-0.06%	High
93	0.32%	Low	0.18%	High
94	0.84%	Low	0.50%	High
95	0.07%	Low	0.25%	High
96	-0.83%	Low	0.15%	High
97	2.14%	Low	0.18%	High

Fig. 11.7 Obtain the data for low- and high-income countries separately.

[Data] tab → [Advanced] in the [Sort and Filter] group. Check that the [List range] is correct, select [Click other location], input **A90** in [Copy to:] and **A85:A86** in [Criteria range] by typing or using the mouse, and click [OK]. The data of the low-income countries are obtained from A85, so check that all incomes are "low". Then change the condition in A86 to **High** and [Copy to] to **C90**, and obtain the data of the high-income countries from C90 (Fig. 11.6, 11.7).

a. Test of *Population* Mean When $\sigma_1^2 = \sigma_2^2 = \sigma^2$

First, we perform the test of the *population* means when $\sigma_1^2 = \sigma_2^2 = \sigma^2$. Generally, it is said that the population growth rate declines as the income increases. So, we consider the null and alternative hypotheses

$$H_0 : \mu_1 = \mu_2,\ H_1 : \mu_1 > \mu_2$$

and set the significance level α to 1%.

Select [Data Analysis] in the [Data] tab. In the menu, select [Analysis Tools], select [t-Test: Two-Sample Assuming Equal Variances] and click [OK] (Fig. 11.8). ([t-Test: Two-Sample Assuming Equal Variances] does not appear in the first screen. So, drag the scroll bar at the right side of the [Analysis Tools] menu. The [t-Test: Two-Sample Assuming Equal Variances] box appears.) Input **A91:A142** (data range of the low-income countries) in [Variable 1 Range] by typing or using the mouse, and **C91:C116** (data range of the high-income countries) in [Variable 2 Range]. The significance level is 1%, so change [Alpha] to **0.01**. Select [Output Range] in [Output Option] and input **F90** (Fig. 11.9). (It is possible

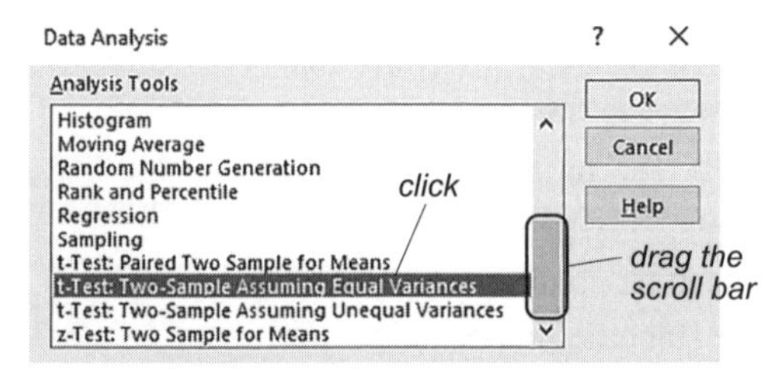

Fig. 11.8 [Data Analysis] in the [Data] tab. In the [Analysis Tools] menu, select [t-Test: Two-Sample Assuming Equal Variances] and click [OK].

to include the field names in [Variable 1 Range] and [Variable 2 Range]; however, it is necessary to click [Labels] in this case.) Click [OK]. The results of the test are displayed from F90 (Fig. 11.10).

The sample means, variances, and observation numbers of the two groups are presented first. [Pooled Variance] is the value calculated by Eq. (11.8). The [Hypothesized Mean Difference] is 0 because the null hypothesis is that the two means are the same, and [df] refers to the degrees of freedom, $n + m - 2$. [t stat] is the test-statistic defined in Eq. (11.9). [P(T<=t) one-tail] is the one-tailed *p*-value, [t Critical one-tail] is the percent point used in one-tailed analysis, [P(T<=t) two-tail] is the two-tailed *p*-value, and [t Critical two-tail] is the percent point used in the two-tailed test.

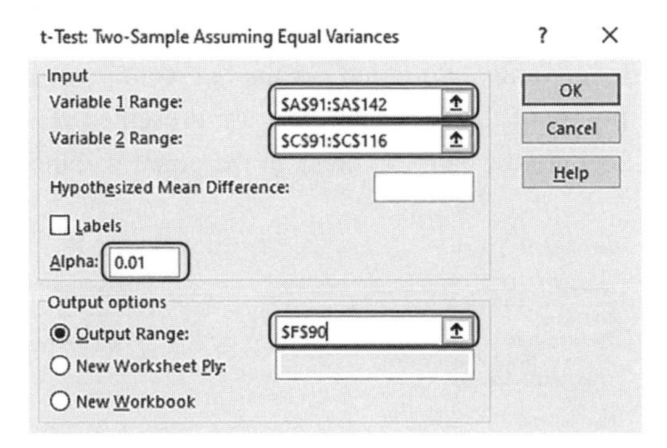

Fig. 11.9 Input **A91:A142** in [Variable 1 Range] and **C91:C116** in [Variable 2 Range]. Change [Alpha] to **0.01**. Select [Output Range] in [Output Option] and input **F90**.

	F	G	H
90	t-Test: Two-Sample Assuming Equal Variances		
91			
92		*Variable 1*	*Variable 2*
93	Mean	0.008259717	0.001804
94	Variance	9.51641E-05	1.82E-05
95	Observations	52	26
96	Pooled Variance	6.98397E-05	
97	Hypothesized Mean Difference	0	
98	df	76	
99	t Stat	3.216221042	
100	P(T<=t) one-tail	0.000954413	
101	t Critical one-tail	2.376420376	
102	P(T<=t) two-tail	0.001908825	
103	t Critical two-tail	2.642078313	

Fig. 11.10 The results of [t-Test: Two-Sample Assuming Equal Variances].

We get $t = 3.216 > t_\alpha(m + n - 2) = 2.376$, and the null hypothesis is rejected. It is admitted that the population growth rate in low-income countries is higher than that in high-income countries.

b. Test of *Population* Mean When $\sigma_1^2 \neq \sigma_2^2$

We do the test of the *population* mean when $\sigma_1^2 \neq \sigma_2^2$. Select the [Data] tab → [Data Analysis] → [t-Test: Two-Sample Assuming Unequal Variances] → [OK] (Fig. 11.11). The [t-Test: Two-Sample Assuming Unequal Variances] box appears. So, as before, input **A91:A142** (data rage of the low-income countries) in [Variable 1 Range] and **C91:C116** (data range of the high-income countries) in [Variable 2 Range]. Change the value of [Alpha] to **0.01**, click [Output Range], input **F110**, and click [OK] (Fig. 11.12).

The sample mean, variance and observation number in each group, difference between the hypotheses, degrees of freedom ν^* obtained from Eq. (11.11), test statistic t, one-tailed p-value, two-tailed p-value, percent point for the one-tailed test, and percent points of the two-tailed test are presented. The expressions of these values in Excel are the same as those in the equal variance case. Since t =

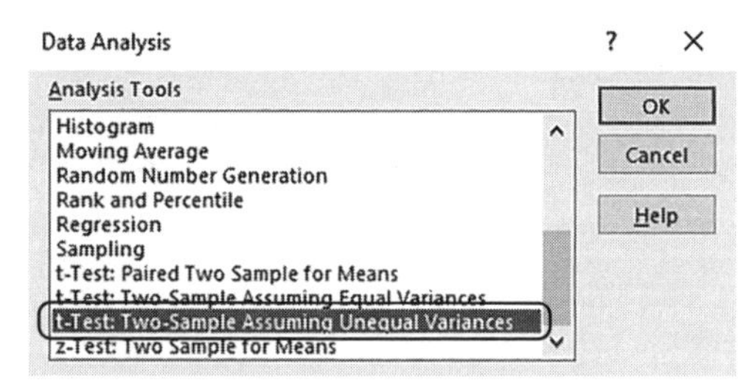

Fig. 11.11 Select [Data] tab → [Data Analysis] → [t-Test: Two-Sample Assuming Unequal Variances] → [OK].

Fig. 11.12 Input [Variable 1 Range], [Variable 2 Range], [Alpha] and [Output Range].

	F	G	H
110	t-Test: Two-Sample Assuming Unequal Variances		
111			
112		*Variable 1*	*Variable 2*
113	Mean	0.008259717	0.001804
114	Variance	9.51641E-05	1.82E-05
115	Observations	52	26
116	Hypothesized Mean Differen	0	
117	df	75	
118	t Stat	4.059403934	
119	P(T<=t) one-tail	5.97973E-05	
120	t Critical one-tail	2.377101812	
121	P(T<=t) two-tail	0.000119595	
122	t Critical two-tail	2.642983067	

Fig. 11.13 Results of [t-Test: Two-Sample Assuming Unequal Variances].

$4.059 > t_\alpha(v^*) = 2.377$, the null hypothesis is rejected as in the equal variance case (Fig 11.13).

c. Test of *Population* Variances

We perform the test of the *population* variances. In this case, we do not have information about the sizes of the variances of the groups, so we do a two tailed-test. The null and alternative hypotheses are

$$H_0 : \sigma_1^2 = \sigma_2^2,\ H_1 : \sigma_1^2 \neq \sigma_2^2$$

Let the significance level α be 5%. Select the [Data] tab → [Data Analysis] → [F-test Two Sample for Variance] → [OK] (Fig. 11.14). The [F-test Two Sample for Variance] box appears. So, input **A91:A142** (data range of the low-income countries) in [Variable 1 Range] and **C91:C116** (data range of the high-income countries) in [Variable 2 Range]. Check that the value of [Alpha] is 0.05 and input **F129** as the [Output Range] (Fig. 11.15). Click [OK]. The results are presented from F129. The sample mean, variance and observation number in each group, degrees of freedom $(m-1, n-1)$, ratio of the two variances F, one-tailed p-value, and percent point for the one-tailed test are presented. However, the percent

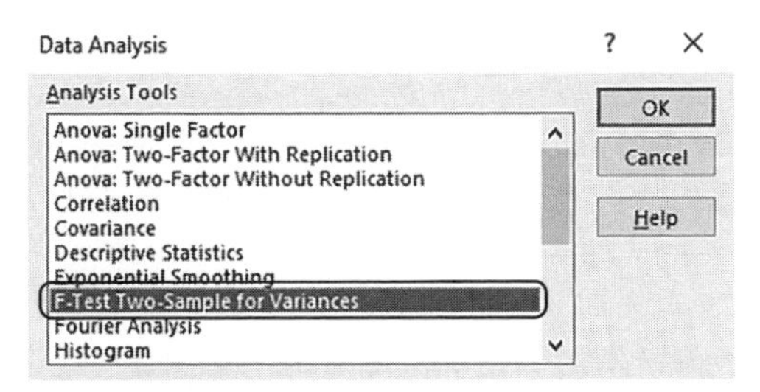

Fig. 11.14 Select [Data] tab → [Data Analysis] → [F-test Two-Sample for Variance] → [OK].

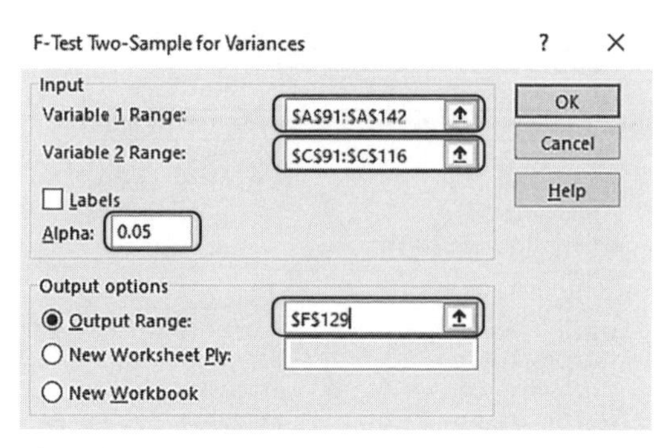

Fig. 11.15 Input [Variable 1 Range], [Variable 2 Range], and [Output Range].

	F	G	H
129	F-Test Two-Sample for Variances		
130			
131		*Variable 1*	*Variable 2*
132	Mean	0.008259717	0.001804
133	Variance	9.51641E-05	1.82E-05
134	Observations	52	26
135	df	51	25
136	F	5.235195244	
137	P(F<=f) one-tail	1.58438E-05	
138	F Critical one-tail	1.839738458	
139			
140			
141	Lower percent point	0.522829466	
142	Upper percent point	2.075563787	

Fig. 11.16 Results of *F*-test for two variances.

points for the two-tailed test are not presented, and we cannot perform the two-tailed test. So, we calculate the percent points using the **FINV** function. Input **Lower percent point**, and **Upper percent point** in F141 and F142, respectively. Input **=FINV(97.5%, 51, 25)** and **=FINV(2.5%, 51, 25)** in G141 and G142, respectively. Since $F = 5.235 > F_{\alpha/2}\ (m - 1, n - 1) = 2.0756$, the null hypothesis is rejected (Fig. 11.16).

In this chapter, we have described two tests of *population* means. As we have already noted, the tests of the *population* mean depend on the equality of the variances. So, in actual data analyses, do the test of the variances first (the significance level is around 5%), and choose the testing method for the *population* means.

11.4 Exercises Using the Population Data

In the exercises, we refer to the [Population Growth Rate] as the growth rate.

1. Perform the point and interval estimations of the *population* mean and variance of the growth rates in high-income countries with a confidence coefficient of 95%.
2. Perform the point and interval estimations of the *population* mean and variance of the growth rate in low-income countries with a confidence coefficient of 99%.
3. Test the null hypothesis, "the *population* mean of the growth rates in high-incomc countries is 0.5%", against the alternative hypothesis, "the mean of the growth rates in high-income countries is not 0.5%", at the 5% significance level.
4. Test the null hypothesis, "the *population* mean of the growth rates in low-income countries is 0.5%", against the alternative hypothesis, "the mean of the growth rates in low-income countries is higher than 0.5%", at the 1% significance level.
5. Test the null hypothesis, "the *population* variance of the growth rates in high-income countries is $(1\%)^2$", against the alternative hypothesis, "the *population* variance of the growth rates in high-income countries is not $(1\%)^2$", at the 5% significance level.
6. Test the null hypothesis, "the *population* variance of the growth rates of low-income countries is $(0.5\%)^2$", against the alternative hypothesis, "the *population* variance of growth rates is larger than $(0.5\%)^2$", at the 1% significance level.
7. Divide the countries into two groups based on the population density:
   High-density countries: countries in which the density is larger than or equal to 100 people/km^2.
   Low-density countries: countries in which the density is less than 100 people/km^2.
   Test "the *population* means of the growth rates in high-density countries and low-density countries are the same" against the alternative hypothesis, "the *population* mean of the growth rates in low-density countries is higher than that in high-density countries", at the 5% significance level, assuming (i) the *population* variances are the same, and (ii) the *population* variances are not the same.
8. In (7), test the null hypothesis "the *population* variances in high- and low-density countries are the same" against the alternative hypothesis "the *population* variances in high- and low-density countries are not the same".

12

Tests of Relationships between Two Variables in Two-Dimensional Data

In Chapter 8, we studied two-dimensional data in which we obtain the values of two variables (x_i, y_i) at the same time for an element i, such as the height and weight of person i. In two-dimensional data, the relationship between two variables becomes very important. In this chapter, we learn methods of testing whether two variables are related or not from the sample (X_1, Y_1), (X_2, Y_2),..., (X_n, Y_n). We learn the χ^2-test of goodness of fit, the one-way layout of the analysis of variance (ANOVA) test, and a correlation coefficient test.

12.1 Test of Independence Using the χ^2-Test of Goodness of Fit

In the χ^2-test of goodness of fit, we compare the expected frequencies obtained from the assumed theorem and the actual observed frequencies. Here, we learn about the general χ^2-test of goodness of fit. Then we learn the independence test using the contingency table.

12.1.1 χ^2-Test of Goodness of Fit

Suppose that n observations are classified into k different categories, $A_1, A_2, \ldots, A_k$, and the observed frequencies in the categories are given by $f_1, f_2, \ldots, f_k$. Let $p_1, p_2, \ldots, p_k$ be the probabilities obtained by the assumed theorem. Then, the expected frequencies, $e_1, e_2, \ldots, e_k$ are given by $e_i = n \cdot p_i$, $i = 1, 2, \ldots, k$. If the assumed theorem is correct, the differences between the observed frequencies and expected frequencies should not be large. Let

(12.1) $$\chi^2 = \sum (f_i - e_i)^2 / e_i$$

Then χ^2 asymptotically follows the χ^2-distribution if the theorem is correct. The number of degrees of freedom of the χ^2 distribution, v, is given by the difference of numbers of unknown parameters between the null and alternative hypotheses. The null hypothesis is that the theorem is correct (alternative hypothesis: the theorem is wrong), and the null hypothesis is rejected when $\chi^2 > \chi^2_{\alpha}(v)$ and accepted otherwise. If the value of χ^2 is small, the observed frequencies match the expected frequencies well; therefore, the rejection region is only the right tail of the χ^2 distribution.

Table 12.1 gives the observed and expected frequencies (the null hypothesis is that all probabilities of obtaining each of the numbers from 1 to 6 are equal to 1/6). Let the significance level be 5%.

Table 12.1 Observed and expected frequencies: Results of 60 trials of rolling a dice.

Number on Dice	1	2	3	4	5	6
Observed Frequencies	11	16	11	7	8	7
Expected Frequencies	10	10	10	10	10	10
Difference	1	6	1	−3	−2	−3

From this table, we get

$$\chi^2 = (1)^2/10 + (6)^2/10 + (1)^2/10 + (-3)^2/10 + (-2)^2/10 + (-3)^2/10 = 6.0$$

If the dice is made correctly, the probabilities of obtaining each of the numbers from 1 to 6 are all 1/6, and there are no unknown parameters. If the dice is not correctly made, we have to estimate the probabilities for all numbers. However, the sum of probabilities is 1, so we need to estimate 5 parameters. Hence the number of degrees of freedom becomes $v = 5$. Since $\chi^2 = 6.0 < \chi^2_{5\%}(5) = 11.071$, we accept the null hypothesis that the dice is correctly made.

12.1.2 Test of Independence Using the Contingency Table

We can test the independence of two variables, X and Y, using the contingency table. Suppose that X takes s different categories, $A_1, A_2, \ldots, A_s$, and Y takes t different categories, $B_1, B_2, \ldots, B_t$. Let f_{ij} be the frequency in (A_i, B_j), $f_{i\bullet} = \sum_{j=1}^{t} f_{ij}$, and $f_{\bullet j} = \sum_{i=1}^{s} f_{ij}$. $f_{i\bullet}, f_{\bullet j}$ are called marginal frequencies and they are the the frequencies of $X = A_i$ and $Y = B_j$. Let $p_{ij} = P(X = A_i, Y = B_j)$, $p_{i\bullet} = P(X = A_i) = \sum_{j=1}^{t} p_{ij}$, and $p_{\bullet j} = P(Y = B_j) = \sum_{i=1}^{s} p_{ij}$. p_{ij} is called the joint probability, and $p_{i\bullet}, p_{\bullet j}$ are called the marginal probabilities.

If X and Y are independent; that is, the occurrence of one does not affect the probability of the other, the null hypothesis becomes

$$H_0 : p_{ij} = p_{i\bullet}p_{\bullet j} \quad \text{for all } i, j$$

The alternative hypothesis is that they are not independent and there is some form of relation between the two variables. $p_{i\bullet}$, $p_{\bullet j}$ are estimated by

(12.2) $$\hat{p}_{i\bullet} = f_{i\bullet}/n, \ \hat{p}_{\bullet j} = f_{\bullet j}/n$$

When the null hypothesis is correct, i.e., when X and Y are independent, the expected frequencies are

(12.3) $$e_{ij} = n\hat{p}_{i\bullet}\hat{p}_{\bullet j} = f_{i\bullet}f_{\bullet j}/n$$

Applying the principle of the goodness-of-fit test, we get

(12.4) $$\chi^2 = \sum_{i=1}^{s}\sum_{j=1}^{t}(f_{ij} - e_{ij})^2/e_{ij}$$

When the variables are independent, the parameters needed to estimate them are the marginal probabilities, $p_{1\bullet}, p_{2\bullet},\ldots, p_{s\bullet}$ and $p_{\bullet 1}, p_{\bullet 2},\ldots, p_{\bullet t}$. The sums of the marginal probabilities for each variable are one, we need to estimate $s + t - 2$ parameters. If the variables are not independent, we need to estimate all p_{ij}; however, the total of the probabilities is also one, and the number of unknown parameters is $s{\cdot}t - 1$.

Therefore, the number of degrees of freedom of the χ^2 distribution is given by $\nu = (s{\cdot}t - 1) - (s + t - 2) = (s - 1){\cdot}(t - 1)$. The test is done by rejecting the null hypothesis (that the two variables are independent) if $\chi^2 > \chi^2_{\alpha}\{(s - 1){\cdot}(t - 1)\}$ and accepting it otherwise.

12.1.3 Test of Independence Using the Population Data

Let's test the independence of [Income] and [Growth Rate] using the contingency table made in Chapter 8. Open the file with the Countries' Population Data, insert [Sheet6] and copy the contingency table (including the title) on [Sheet4] beginning at A1 of [Sheet6] (use [Paste Values]). We calculate the expected frequencies when the variables are independent. Input **Table 2 Expected Frequencies: Independent Case** in A11, **Growth Rate** in C12, **Income** in A13, **High** in A14, **Low** in A15, **High** in B13, **Middle** in C13, and **Low** in D13. Input **=$E4*B$6/$E$6** in B14 and copy it to the entire table range (B14:D15). We combine the relative and absolute cell addresses, so we do not have to input the equation more than once (Fig. 12.1).

The expected frequencies are calculated, and we obtain the relative error $(f_{ij} - e_{ij})^2/e_{ij}$. Input **Table 3 Relative Errors** in A20. From A21, input the variable and category names so that the format of the table is the same as that of Table 2. Input **=(B4–B14)^2/B14** in B23 and copy it to the entire table (B23:D24). From A27 to

	A	B	C	D	E
1	Table 1 Contingency Table of Income and Population Growth Rate				
2	Sum of Frequency	Column Labels			
3	Row Labels	High	Middle	Low	Grand Total
4	High	1	4	21	26
5	Low	20	9	23	52
6	Grand Total	21	13	44	78
7					
8					
9					
10					
11	Table 2 Expected Frequencies: Independent Case				
12		Growth Rate			
13	Income	High	Middle	Low	
14	High	7.000	4.333	14.667	
15	Low	14.000	8.667	29.333	

Fig. 12.1 Observed and expected (independent case) frequencies.

A30, input **Test Statistic chi2**, **Degrees of Freedom**, **Significance Level**, and **Percent Point**. Input **=SUM(B23:D24)** and calculate the sum of relative errors in B27. From B28 to B30, input the degrees of freedom = $(s-1)(t-1)$, **2** in B28, input the significance level **5%** in B29, and calculate the percent point using **=CHIINV(B29,B28)** in B30. Since $\chi^2 = 11.855 > \chi^2_\alpha\{(s-1)(t-1)\} = 5.991$, the null hypothesis is rejected and some form of relation between the variables is admitted (Fig. 12.2).

	A	B	C	D
20	Table 3 Relative Errors			
21		Growth Rate		
22	Income	High	Middle	Low
23	High	5.143	2.735	0.026
24	Low	2.571	1.367	0.013
25				
26				
27	Test Staistic chi2	11.855		
28	Degrees of Freedom	2		
29	Significance Level	5%		
30	Percent Point	5.991		
31	chi2>percent point			
32	Reject Null Hypothesis			

Fig. 12.2 Results of the test of independence.

Excel has the **CHITEST** function to perform the goodness-of-fit test. It is used as follows: **CHITEST(**range of observed frequencies, range of expected frequencies). This function calculates the one-tailed p-value of the χ^2 distribution. We reject the null hypothesis if the value of the function is smaller than the significance level α, and accept it otherwise. However, we cannot input the degrees of freedom. The number of degrees of freedom becomes: (i) $k-1$ if the data are one row or one column where k is the number of categories, or (ii) $(s-1)\cdot(t-1)$ if data consist of s rows and t columns ($s, t \geq 2$). So, we can use this function when all theoretical probabilities are given or when testing the independence of two variables. However, we may not use it in other cases.

12.2 One-Way Layout ANOVA

Suppose there are s different normal *populations* and they follow $N(\mu_1, \sigma^2)$, $N(\mu_2, \sigma^2)$, …, $N(\mu_s, \sigma^2)$. To analyze *population* means of 3 or more, we use the analysis of variance (ANOVA). One possibility is that there exists a certain variable A that may affect the *population* means. It is important that the effects of A differ by *population*. They are classified into s different categories, $A_1, A_2, \ldots, A_s$. For example, we set certain conditions for experiments or observations. The variable that may affect the results is called a factor, and the different categories of the factor are called levels. When there is just one factor, we call it a one-way layout, and when multiple factors affect the results, we call it a multi-way layout. In this section, we learn about the one-way layout model of ANOVA.

12.2.1 One-Way Layout Model

Suppose that the levels of the factor A are $A_1, A_2, \ldots, A_s$ and we have $n_1, n_2, \ldots, n_s$ observations of the levels. Let the j-th observation of the level i be Y_{ij}. We assume that the mean differs depending on the level but the variances are the same independent of the levels, and Y_{ij} follows the normal distribution with mean μ_i and variance σ^2, i.e., $N(\mu_i, \sigma^2)$. (For accuracy, μ_i is the expected value of Y_{ij} or the *population* mean at the level i, but we generally use these notations.) Let the number of total observations be $n = \sum_{i=1}^{s} n_i$ and let the weighted average (the weight is based on the number observations of each level) be

$$\mu = \sum_{i=1}^{s} n_i \mu_i / n \quad (12.5)$$

μ is called the grand mean, and

$$\delta_i = \mu_i - \mu \quad (12.6)$$

is called the effect of A_i. Note that $\sum n_i \delta_i = 0$.

12.2.2 ANOVA

Next, we analyze the data of the one-way layout model by ANOVA. The null hypothesis is that the means are the same at all levels and there are no effects of the levels, and it is given by

$$H_0 : \mu_1 = \mu_2 = \cdots = \mu_s = \mu \quad \text{or} \quad H_0 : \delta_1 = \delta_2 = \cdots = \delta_s = 0$$

The alternative hypothesis is H_1: $\mu_i \neq \mu$ or $\delta_i \neq 0$ for at least one level i.

Now, estimate μ using all observations by

(12.7) $$\bar{Y}_{\bullet\bullet} = \sum_{i=1}^{s}\sum_{j=1}^{t} Y_{ij}/n$$

μ_i is estimated using the sample mean of each level,

(12.8) $$\bar{Y}_{i\bullet} = \sum_{j=1}^{n_i} Y_{ij}/n_i$$

Let the sums of squared deviations from $\bar{Y}_{\bullet\bullet}$, $\bar{Y}_{i\bullet}$ be

(12.9)
$$S_t = \sum_{i=1}^{s}\sum_{j=1}^{v}(Y_{ij} - \bar{Y}_{\bullet\bullet})^2$$
$$S_e = \sum_{i=1}^{s}\sum_{j=1}^{t}(Y_{ij} - \bar{Y}_{i\bullet})^2 = \sum_{i=1}^{s} n_i(Y_{ij} - \bar{Y}_{i\bullet})^2$$

S_t, S_e are called the total variation and the within-class variation, respectively. It is known that S_e/σ^2 follows a χ^2 distribution with the degrees of freedom given by $v_e = n - s$. Let

(12.10) $$S_a = S_t - S_e = \sum_{i=1}^{s} n_i(\bar{Y}_{i\bullet} - \bar{Y}_{\bullet\bullet})^2$$

S_a is called the between-class variation. If the null hypothesis is correct, $\bar{Y}_{i\bullet} \approx \bar{Y}_{\bullet\bullet}$ for all i, and S_a cannot be a large value. In this case, S_a/σ^2 and S_e/σ^2 are independent and follows the χ^2 distribution with the degrees of freedom given by $v_a = s - 1$.

Therefore, under the null hypothesis,

(12.11) $$F = \frac{S_a/v_a}{S_e/v_e}$$

follows the F-distribution with the degrees of freedom given by (v_a, v_e), $F(v_a, v_b)$. Using this relation, we compare F and the percent point of the F-distribution, $F_\alpha(v_a, v_e)$, and we reject the null hypothesis if $F > F_\alpha(v_a, v_e)$ and accept it otherwise. This test is called the ANOVA.

Like the goodness-of-fit test, the rejection region is just the right tail. When $F = 0$, $S_a = \sum_{i=1}^{s} n_i(\bar{Y}_{i\bullet} - \bar{Y}_{\bullet\bullet})^2 = 0 \Leftrightarrow \bar{Y}_{1\bullet} = \bar{Y}_{2\bullet} = \cdots = \bar{Y}_{s\bullet} = \bar{Y}_{\bullet\bullet}$, and the means of all levels are the same, and there is no reason to reject the null hypothesis. The p-values are defined by the probability larger than F in $F(v_a, v_e)$.

When $s = 2$, the test statistic t of the two-sample test under equivariances in the previous chapter satisfies $t^2 = F$, and the result of ANOVA is the same as that of the two-tailed test in the two-sample test when the variances are the same. Since we cannot use the one-tailed test or Welch's test in ANOVA, we should use the t-test described in the previous chapter when $s = 2$.

12.2.3 One-Way ANOVA Using the Population Data

Let's analyze whether the region affects the population growth rate using the Countries' Population Data. Excel has a procedure for conducting one-way layout ANOVA in [Data Analysis]. We analyze the data of three regions, Africa, Asia, and Europe.

First, we divide the data by region. Copy the [Region] and [Population Growth Rate] data in [Sheet1] to the range from H1 in [Sheet6] (for [Population Growth Rate], use [Paste Values]). Next, copy the field name [Region] to K1 and input **Africa** in K2. Since we just need the data for the [Population Growth Rate], copy the field name [Population Growth Rate] to M1. Move the active cell in the list and click [Data] → [Advanced] in the [Sort & Filter] group. Check to ensure that the [List range] is correct, input **K1:K2** in [Criteria range], click [Copy to another location], input **M1** in the [Copy to], and click [OK]. The [Population Growth Rate] data for Africa are extracted. Change M1 to **Africa** so that we can specify the region. In the same way, change the criterion in K2 and extract the [Population Growth Rate] for Asia from N1 and for Europe from O1 (Fig. 12.3).

For the one-way ANOVA test, the data must be in consecutive columns as in this example. Let the significance level be 5%. Click the [Data] tab → [Data Analysis]. Select [Anova: Single Factor] in the menu and click [OK] (Fig. 12.4).

The [Anova: Single Factor] box appears. Input **M2:O23** in [Input Range] by typing or using the mouse and check to ensure that the value of [Alpha] is 0.05. Click [Output Range] and input **M30** in [Output Range] (Fig. 12.5). Click [OK]. The results are presented from M30 (Fig. 12.6).

We can include the field names in [Input Range] and make the range M1:O23, but click [Labels in the first row]; otherwise an error occurs. (In this case, the length of the columns is 22, which is longer than that of the rows, which is 3. The data are treated as column data if the length of the column is larger, and as row

	M	N	O
1	Africa	Asia	Europe
2	0.71%	0.32%	-0.06%
3	2.14%	-0.33%	0.18%
4	1.95%	0.37%	-0.83%
5	2.31%	0.35%	0.25%
6	2.06%	0.08%	0.15%
7	1.11%	1.07%	0.18%
8	1.43%	-0.54%	-0.19%

Fig. 12.3 Obtain [Population Growth Rate] of Africa, Asia, and Europe.

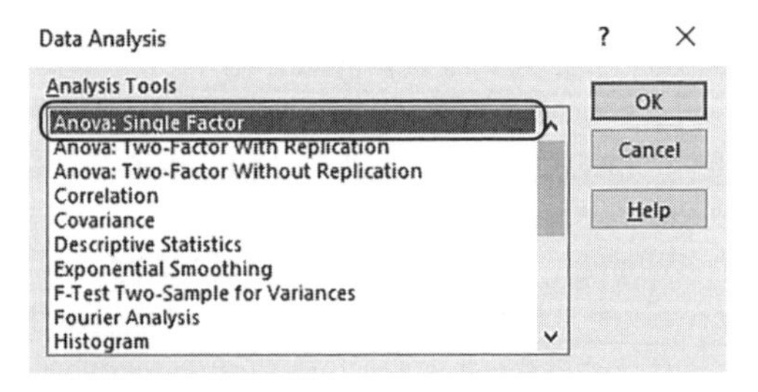

Fig. 12.4 Click [Data] tab → [Data Analysis]. Select [Anova: Single Factor] in the menu and click [OK].

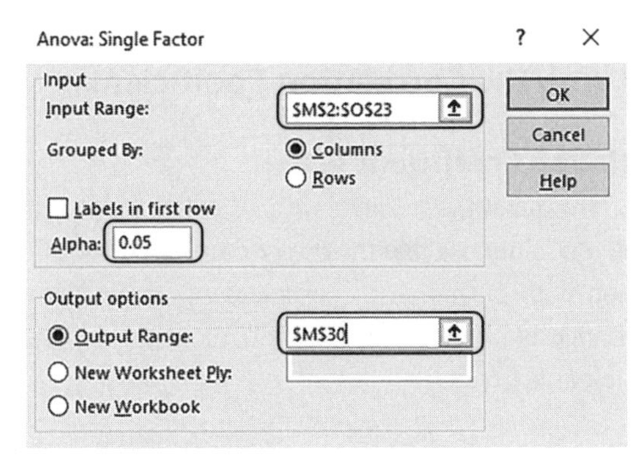

Fig. 12.5 Input **M2:O23** in [Input Range], check to ensure that the value of [Alpha] is 0.05, and input **M30** in [Output Range].

	M	N	O	P	Q	R	S
30	Anova: Single Factor						
31							
32	SUMMARY						
33	*Groups*	*Count*	*Sum*	*Average*	*Variance*		
34	Column 1	21	0.362756777	0.017274	5.5E-05		
35	Column 2	21	0.067718834	0.003225	2.38E-05		
36	Column 3	22	-0.025596935	-0.001163	1.81E-05		
37							
38							
39	ANOVA						
40	*Source of Variation*	*SS*	*df*	*MS*	*F*	*P-value*	*F crit*
41	Between Groups	0.003953	2	0.001977	61.68634	2.23262E-15	3.147791
42	Within Groups	0.001955	61	3.2E-05			
43							
44	Total	0.005908	63				
45							

Fig. 12.6 Results of one-way layout ANOVA.

data if the length of the row is larger than that of the column. So, if s is larger than the maximum number of observations in the levels, set [Grouped by] to [Column].)

The summaries of three groups and the ANOVA table are presented. From the ANOVA table, we get a between-class variation (*ss* of [Between Groups] in Excel) S_a = 0.003953, a within-class variation (*ss* of [Within Groups]) S_e = 0.001955, degrees of freedom (*df*) $\nu_a = 2$, $\nu_e = 61$, a test statistic (F of [Between Groups]) $F = 61.686$, and a percent point (F crit of [Within Groups]) $F_\alpha(\nu_a, \nu_e)$ = 3,148. Therefore, $F > F_\alpha(\nu_a, \nu_e)$ and the null hypothesis is rejected; it is admitted that the regions affect the population growth rates. Note that we can compare the [P-value] of [Between Groups] and significance level α. Since the p-value $< \alpha$, we reject the null hypothesis.

12.3 Tests Using the Correlation Coefficient

12.3.1 Correlation Coefficient Tests

When X and Y are quantitative data, we can calculate the (product-moment) correlation coefficient. Suppose that the *population* of (X, Y) follows the bivariate normal distribution with means μ_x, μ_y variances σ_x^2, σ_y^2, and covariance σ_{xy}. The *population* covariance is obtained by $\sigma_{xy} = E[(X - \mu_x)(Y - \mu_y)]$. The *population* correlation coefficient is given by

$$\rho = \sigma_{xy}/(\sigma_x \sigma_y) \tag{12.12}$$

As we have studied in Chapter 8, ρ satisfies $-1 \leq \rho \leq 1$ and expresses the linear relationship between two variables, and

i) $\rho = \pm 1$: the variables satisfy $Y = a + bX$, $b \neq 0$, and $\rho = 1$ if $b > 0$ and $\rho = -1$ if $b < 0$.

ii) $\rho > 0$: positive correlation, Y tends to increase as X increases, and the relation becomes stronger as ρ approaches 1.

iii) $\rho < 0$: Y tends to decrease as X increases, and the relation becomes stronger as ρ approaches -1.

iv) $\rho = 0$: no linear relations between two variables, and they are uncorrelated.

In general, "uncorrelated" means that there is not a linear relation between two variables, but it does not mean that there is no relation between two variables. Suppose X follows the standard normal distribution, and $Y = X^2$. It is obvious that there is a strong relation between X and Y. But we get $\rho = 0$. When X and Y are independent, they are uncorrelated, but the variables may not be independent when X and Y are uncorrelated.

However, if (X, Y) follows the bivariate normal distribution as in this section, the variables become independent if they are uncorrelated.

When we get (X_1, Y_1), (X_2, Y_2),..., (X_n, Y_n) as the sample, the sample covariance and correlation coefficient are obtained by

$$s_{xy} = \sum (X_i - \bar{x})(Y_i - \bar{y})/(n-1)$$

$$r = \frac{s_{xy}}{s_x s_y} = \frac{\Sigma (X_i - \bar{x})(Y_i - \bar{y})}{\sqrt{\Sigma (X_i - \bar{x})^2}\sqrt{\Sigma (Y_i - \bar{y})^2}} \tag{12.13}$$

where s_x and s_y are the sample standard deviations of X and Y.

Now, let the null hypothesis be

$$H_0 : \rho = \rho_0$$

We use different methods whether $\rho_0 = 0$ or not. When $\rho_0 = 0$, the null hypothesis becomes H_0: $\rho = 0$ and

$$t = r\sqrt{n-2}/\sqrt{1-r^2} \tag{12.14}$$

follows the t-distribution with degrees of freedom $n-2$, $t(n-2)$, under the null hypothesis. We test using this relation. First calculate the test statistic t from Eq. (12.14) and obtain the percent values of $t_\alpha(n-2)$ and $t_{\alpha/2}(n-2)$. Then:

i) When, $H_1: \rho \neq 0$, we do a two-tailed test and reject H_0 if $|t| > t_{\alpha/2}(n-2)$ and accept it otherwise.

ii) When $H_1: \rho > 0$, we do a right one-tailed test and reject H_0 if $t > t_\alpha(n-2)$ and accept it otherwise.

iii) When $H_1: \rho < 0$, we do a left one-tailed test and reject H_0 if $t < -t_\alpha(n-2)$ and accept it otherwise.

When $\rho \neq 0$, the distribution of r is very complicated. Therefore, we usually use the approximation method called Fisher's z-transformation. Let

$$z = \frac{1}{2}\log_e\{(1+r)/(1-r)\},\ \eta = \frac{1}{2}\log_e\{(1+\rho)/(1-\rho)\}, \tag{12.15}$$

Then $Z = \sqrt{(n-3)}(z-\eta)$ approximately follows the standard normal distribution, $N(0, 1)$. We use $n-3$ to get a better approximation. Using this relation, we perform the test. We substitute the value of ρ_0 for ρ, calculate the value of Z, and compare it to the percent points of the standard normal distribution, Z_α, $Z_{\alpha/2}$.

12.3.2 Test of the Correlation Coefficient Using the Countries' Population Data

Here, we test the relation between the [Log of GDP] and the [Population Growth Rate] using the correlation coefficient. It is said that the population growth rate decreases as the income increases. We set

$$H_0 : \rho = 0, \quad H_1 : \rho < 0$$

and set the significance level α at 5%. Since H_0: $\rho = 0$, we use Eq. (12.14).

Input **Population Growth Rate and Log of GDP per capita** and **Sample Correlation Coefficient** in R1 and R2 in [Sheet6]. The correlation coefficient was already calculated in Chapter 8; copy it from [Sheet4] to S2 (use [Paste Values]). (We have divided by n to calculate the standard deviations and covariance. When we calculate the correlation coefficient, we get the correct result if we divide the denominator and numerator by the same number.)

From R3 to R7, input **Sample Size(n), Significance Level, Degrees of Freedom(n-2), Test Statistic t,** and **Percent Points**. Input **78** in S3, **1%** in S4, and **=S3-2** in S5. Input **=SQRT(S5)*S2/SQRT(1−S2^2)** in S6 and **=TINV(2*S4, S5)** and obtain the test statistic t and the percent point.

Since $t = -7.955 < -t_\alpha(n-2) = -2.376$, the null hypothesis is rejected, and a

	R	S
1	Population Growth Rate and Log of GDP per capita	
2	Sample Correlation Coeffcient	-0.67407
3	Sample Size (n)	78
4	Significance Level	1%
5	Degrees of Freedom(n-2)	76
6	Test Statistic t	-7.95535
7	Percent Point	2.37642
8	t<-Percent Point	
9	Reject Null Hypothesis	

Fig. 12.7 Results of the test using the correlation coefficient.

negative correlation between the income and population growth rate is admitted (Fig. 12.7).

12.4 Exercises Using the Population Data

1. Using the contingency table of the population density and population growth rate made in the exercises in Chapter 8, test the independence of the two variables using the χ^2 test of goodness of fit at the 5% significance level.
2. Divide the countries into four different groups based on regions (Africa, Asia, Europe, and other regions), and test whether the region affects the population growth or not using one-way ANOVA at the 1% significance level.
3. Using the correlation coefficient of the log of the population density and the population growth rate, test H_0: $\rho = 0$, H_1: $\rho \neq 0$ at the 5% significance level. As an exercise, do the tests using both Eq. (12.14) and (12.15) based on Fisher's z-transformation. The percent point of the normal distribution is obtained by the **NORMSINV** function; however, unlike the t, χ^2, and F-distributions, **NORMSINV(p)** calculates the point below which the probability becomes p. So, it is necessary to obtain Z_α using **NORMSINV(1−*α*)**.

13

Regression Analysis

Suppose there are two-dimensional variables (X, Y). In the regression analysis, we estimate the regression equation in which Y is quantitatively explained by X. Y is called the dependent (or explained) variable and X is called the independent (or explanatory) variable. We assume that we know the causal relationship between X and Y; that is, X is the cause and Y is the result.

Regression analysis is a very important, core method in statistical analysis. The word "regression" was first used by Francis Galton (1822–1911). He analyzed the heights of fathers and their sons. He found that the son was tall if the father was tall. But he found a tendency toward the mean, that is that tall and short fathers tended to have sons whose heights were closer to the average height. He described this tendency as the heights of the sons regressing to the mean. Nowadays, the meaning of "regression" has been generalized to describe the quantitative relationships between two or more variables.

13.1 Simple Regression Analysis

We first discuss a model that contains only one independent variable. This is called simple regression analysis.

13.1.1 Linear Regression Model

As we have studied in Chapters 8 and 12, the income level (log of GDP per capita) and population growth rate are related. It can be observed that the population growth rate tends to decrease as the income level increases. But this analysis is qualitative. We want to determine the quantitative relations among these variables in order to help solve the world's population problems. Let X be the log of GDP per capita and let Y be the population growth rate. X is the independent (or explanatory) variable and Y is the dependent (or explained) variable in this case. As shown in the scatter diagram in Fig. 8.4:

i) The values of Y tend to decrease as the values of X decrease.

ii) The values of Y vary for the similar values of X.

Therefore, we can divide Y into two parts: a part that changes systematically with X and a part that changes randomly. Suppose the first part is given by a linear function of x:

(13.1) $$y = \beta_1 + \beta_2 x.$$

This is called the regression equation or regression function. In this chapter we consider the case where y is a linear function of x. This is called the linear regression. Note that even if the original equation is nonlinear, we can obtain the linear model by transforming functions such as by taking the log in many cases. We can also approximate the model with the linear model using the Taylor expansion. The linear model can be used in various cases, and it is a core model in statistical models.

Let the i-th observation be (X_i, Y_i) and the randomly changing part be u_i. Then we can consider the model

(13.2) $$Y_i = \beta_1 + \beta_2 X_i + u_i, \ i = 1, 2, \ldots, n.$$

This model is called the *population* regression equation. β_1, β_2 are unknown parameters called the *population* (partial) regression coefficients. u_i is called the error term.

13.1.2 Assumptions

X_i and u_i satisfy the following assumptions.

Assumption 1

X_i is nonstochastic and takes fixed values.

Assumption 2

u_i is a random variable and $E(u_i) = 0, \ i = 1, 2, \ldots, n$

Assumption 3

Error terms are uncorrelated with each other, i.e., if $i \neq j$, $Cov(u_i, u_j) = E(u_i u_j) = 0$

Assumption 4

Variances of error terms are the same and σ^2, i.e., $V(u_i) = E(u_i^2) = \sigma^2, \ i = 1, 2, \ldots, n$

Under these assumptions, the expected value of Y_i becomes

(13.3) $$E(Y_i) = \beta_1 + \beta_2 X_i.$$

13.2 Least Squares Method

13.2.1 Estimation of the Regression Coefficients

β_1, β_2 are unknown parameters. It is necessary for us to estimate them using the obtained data, (X_1, Y_1), (X_2, Y_2),…, (X_n, Y_n). The portion of Y_1 which cannot be explained by X_1 is $u_i = Y_i - (\beta_1 - \beta_2 X_i)$.

To remove the effects of signs, we square it and consider their sum,

$$S = \sum u_i^2 = \sum \{Y_i - (\beta_1 + \beta_2 X_i)\}^2.$$

Since S is the portion that cannot be explained, it is desirable to make it as small as possible. The method that minimizes S and gets $\hat{\beta}_1$ and $\hat{\beta}_2$, which are estimators of β_1 and β_2, is called the least squares method. $\hat{\beta}_1$ and $\hat{\beta}_2$ are called the least squares estimators.

We can obtain $\hat{\beta}_1$ and $\hat{\beta}_2$ by partially differentiating S and put 0,

$$\frac{\partial S}{\partial \beta_1} = -2 \sum (Y_i - \beta_1 - \beta_2 X_i) = 0$$

$$\frac{\partial S}{\partial \beta_2} = -2 \sum (Y_i - \beta_1 - \beta_2 X_i) X_i = 0.$$

Solving these equations, we get

$$\hat{\beta}_2 = \frac{\Sigma (X_i - \bar{X})(Y_i - \bar{Y})}{\Sigma (X_i - \bar{X})^2} \tag{13.4}$$

$$\hat{\beta}_1 = \bar{Y} - \hat{\beta}_2 \bar{X}.$$

$\bar{X}$ and $\bar{Y}$ are the sample means of X and Y. $\hat{\beta}_1$ and $\hat{\beta}_2$ are called sample (partial) regression coefficients, $y = \hat{\beta}_1 + \hat{\beta}_2 x$ is called the sample regression equation or sample regression line.

The estimator of $E(Y_i)$ using the sample regression equations is called the regression value or fitted value and is given by

$$\hat{Y}_i = \hat{\beta}_1 + \hat{\beta}_2 X_i. \tag{13.5}$$

Then

$$e_i = Y_i - \hat{Y}_i \tag{13.6}$$

is the portion that cannot be explained by X_i. This is called the residual. e_i is an estimator of u_i and called a residual. We always get

$$\sum e_i = 0, \ \sum e_i X_i = 0.$$

(The first equation corresponds to $\partial S/\partial \beta_1 = 0$ and the second corresponds to $\partial S/\partial \beta_2 = 0$.)

The variance of u_i, σ^2, is estimated as

$$s^2 = \sum e_i^2/(n-2). \tag{13.7}$$

We divide by $(n - 2)$, because there are two constraints on e_i, and the number of degrees of freedom is reduced by 2. The value we obtain when we substitute the observed values into the estimator is called the estimate.

13.2.2 Properties of the Least Squares Estimator

For the estimation of $\hat{\beta}_1$ and $\hat{\beta}_2$, we use the least squares method. However, to eliminate the effect of the sign of $u_i = Y_i - (\beta_1 - \beta_2 X_i)$, we may consider the sum of the absolute values and the estimator that minimizes,

$$D = \sum |u_i| = \sum |Y_i - (\beta_1 - \beta_2 X_i)|.$$

This method is called the least absolute deviations method, and the estimator is called the least absolute deviations estimator. The reason of using the least squares method is that we consider the expected value and mean as the location parameter and adopt various assumptions regarding u_i, $E(u_i) = 0$, etc. Under these assumptions, the least squares estimator has superior properties that other estimators do not have. (The least absolute deviations method corresponds to the model that considers the median of $u_i = 0$.)

$\hat{\beta}_1$ and $\hat{\beta}_2$ satisfy

$$E(\hat{\beta}_1) = \beta_1,\ E(\hat{\beta}_2) = \beta_2$$

They are unbiased estimators. s^2 also satisfies $E(s^2) = \sigma^2$ and is an unbiased estimator of σ^2. The variances are

$$V(\hat{\beta}_1) = \frac{\sigma^2 \sum X_i^2}{n \sum (X_i - \bar{X})^2}$$
$$V(\hat{\beta}_2) = \frac{\sigma^2}{\sum (X_i - \bar{X})^2} \qquad (13.8)$$

In practice, σ^2 is unknown, we replace it with s^2.

$\hat{\beta}_1$ and $\hat{\beta}_2$ are the best linear unbiased estimators (BLUE) which have the smallest variances among linear unbiased estimators according to Gauss-Markov's theorem. Linear unbiased estimators are given by

$$\tilde{\beta}_j = \sum_i c_{ij} Y_i,\ E(\tilde{\beta}_j) = \beta_j,\ j = 1, 2.$$

Especially if u_i is independent and follows the normal distribution, the least squares estimator becomes the best unbiased estimator according to Cramér-Rao's inequality.

13.2.3 Goodness of Fit and R^2

The goodness of fit of the regression equation, that is how well X explains Y, is an important factor to consider when determining the usefulness of the model. If X can explain a large portion of the variations in Y, the model may be a useful one. If only a small portion of the variations is explained, the model may not be very useful. The criterion used for this purpose is the coefficient of determination, R^2.

The total variation of Y_i is $\sum (Y_i - \bar{Y})^2$. The portion that can be explained by the regression equation is $\sum(\hat{Y}_i - \bar{Y})^2$, the portion that cannot be explained is $\sum e_i^2$, and, we get

$$\sum(Y_i - \bar{Y})^2 = \sum(\hat{Y}_i - \bar{Y})^2 + \sum e_i^2.$$

R^2 is the portion that can be explained and is given by

$$R^2 = 1 - \frac{\sum e_i^2}{\sum(Y_i - \bar{Y})^2} = \frac{\sum(\hat{Y}_i - \bar{Y})^2}{\sum(Y_i - \bar{Y})^2}. \quad (13.9)$$

R^2 satisfies $0 \leq R^2 \leq 1$ and becomes 1 if X_i perfectly explains Y_i. Let r be the correlation coefficient of X_i and Y_i. Then, $R^2 = r^2$.

13.3 Sample Distributions of Regression Coefficients and Hypothesis Testing

In regression analysis, we also perform hypothesis tests of β_1 and β_2. For that purpose, it is necessary for us to know the sample distributions of $\hat{\beta}_1$ and $\hat{\beta}_2$. In addition to the assumptions listed earlier, we add the assumption that $u_1, u_2, \ldots, u_n$ are independent and follow the normal distribution. Even if the distribution of $u_1, u_2, \ldots, u_n$ is not normal, we can asymptotically (i.e., approximately when n is large) use the results in accordance with the central limit theorem.

13.3.1 Sample Distributions of the Estimators

For $\hat{\beta}_2$, we get

$$\hat{\beta}_2 = \beta_2 + \frac{\sum(X_i - \bar{X})u_i}{\sum(X_i - \bar{X})^2}.$$

$\hat{\beta}_2$ is the sum of independent normal random variables. Therefore, the distribution of $\hat{\beta}_2$ is

$$N\left(\beta_2, \frac{\sigma^2}{\sum(X_i - \bar{X})^2}\right)$$

$(\hat{\beta}_2 - \beta_2)\sigma\Big/\sqrt{\sum(X_i - \bar{X})^2}$ follows the standard normal distribution. The "standard error" of $\hat{\beta}_2$ is obtained by

$$s.e.(\hat{\beta}_2) = \frac{s}{\sqrt{\sum(X_i - \bar{X})^2}} \quad (13.10)$$

where σ is replaced by s in $(\hat{\beta}_2 - \beta_2)\sigma/\sqrt{\sum(X_i - \bar{X})^2}$. It is known that

$$t_2 = (\hat{\beta}_2 - \beta_2)/s.e.(\hat{\beta}_2) \quad (13.11)$$

follows the t distribution with the degrees of freedom given by $n - 2$, $t\ (n - 2)$. Note that this test gives the same result as the correlation coefficient test studied in a previous chapter.

The distribution of $\hat{\beta}_1$ is

$$N\left(\beta_1, \frac{\sigma^2 \sum X_i^2}{n\sum(X_i - \bar{X})^2}\right).$$

The standard error is calculated by

$$s.e.(\hat{\beta}_1) = s\sqrt{\frac{\sum X_i^2}{n\sum(X_i - \bar{X})^2}} \tag{13.12}$$

and

$$t_1 = (\hat{\beta}_1 - \beta_1)/s.e.(\hat{\beta}_1) \tag{13.13}$$

follows $t\,(n-2)$ as the $\hat{\beta}_2$ case.

13.3.2 Hypothesis Tests of Regression Coefficients

Let's consider the hypothesis test of β_2. Let the null hypothesis be

$$H_0 : \beta_2 = a.$$

The alternative hypothesis is H_1: $\beta_2 \neq a$ (two-tailed), H_1: $\beta_2 > a$ (right one-tailed), or H_1: $\beta_2 < a$ (left one-tailed).

We choose the type of alternative hypothesis based on the prior information and the purpose of the test.

First, we calculate $\hat{\beta}_2$ and $s.e.(\hat{\beta}_2)$, and obtain the value of the test statistic

$$t_2 = (\hat{\beta}_2 - a)/s.e.(\hat{\beta}_2). \tag{13.14}$$

The critical value, which determines the rejection region, is obtained using the percent point of the t-distribution with the degrees of freedom given by $n-2$, $t_{\alpha/2}(n-2)$ or $t_\alpha(n-2)$. The test is done by comparing t_2 and $t_{\alpha/2}(n-2)$ or $t_\alpha(n-2)$.

i) If H_1: $\beta_2 \neq a$, we reject the null hypothesis if $|t_2| > t_{\alpha/2}(n-2)$ and accept it otherwise.

ii) If H_1: $\beta_2 > a$ we reject the null hypothesis if $t_2 > t_\alpha(n-2)$ and accept it otherwise.

iii) If H_1: $\beta_2 < a$ we reject the null hypothesis if $t_2 < -t_\alpha(n-2)$ and accept it otherwise.

Regression analysis is a method of explaining Y by X. Therefore, the test of whether X can explain Y, i.e., the test of H_0: $\beta_2 = 0$, becomes important. If this null hypothesis is rejected, we say the regression equation is significant. (The coefficient is significantly different from zero.) $t_2 = \hat{\beta}_2/s.e.(\hat{\beta}_2)$ is called the t-ratio or t-value.

Although it is not as common as the β_2 case, we can also test β_1 using t_1.

13.4 Simple Regression Analysis Using the Population Data

13.4.1 Estimation of the Model

In this chapter, we perform a regression analysis using the Countries' Population Data. Insert [Sheet7] and copy the [Population Growth Rate], [GDP per capita], [Population Density], and [Region] from A1 in this order (use [Paste Values] for [Population Growth Rate] and [Population Density]. We use [Population Density] and [Region] later.)

We consider the model in which Y is the [Population Growth Rate] and X is the logarithm of [GDP per capita]. Unlike in previous cases, we use the natural logarithm whose base is $e = 2.718\ldots$. We cannot explain the meaning of the coefficient unless we use the natural logarithm. Input **Ln of GDP** in E1 **=LN(B2)** in E2 and copy E2 to E3:E79. **Ln** is the function that calculates the natural logarithm. The *population* regression equation becomes

$$\begin{aligned} &Y_i = \beta_1 + \beta_2 X_i + u_i, \quad i = 1, 2, \ldots, n \\ &Y_i = [\text{Population Growth Rate}], \; X_i = [\text{Ln of GDP}] \end{aligned} \quad (13.15)$$

The coefficient β_2 expresses the change in the [Population Growth Rate] when [GDP per capita] increases by 1%. When we do a transformation of a variable, the meaning of the coefficient changes. Now, some software packages automatically choose the best-fitting model. However, we have to understand the meaning of coefficients when transformations of variables are performed.

As we have already studied, estimation of the regression equation requires slightly complicated calculations. Excel has a procedure for performing regression analysis. Select [Data] tab → [Data Analysis] → [Regression] (Fig. 13.1). The [Regression] box opens, and we input the data ranges. Input **A1:A79** in [Input Y Range] and **E1:E79** in [Input X Range] by typing or using the mouse. Since the data ranges include the field names, click the [Labels] so that the first row is the field name (otherwise an error occurs). Click [Output Range] in [Output Option]

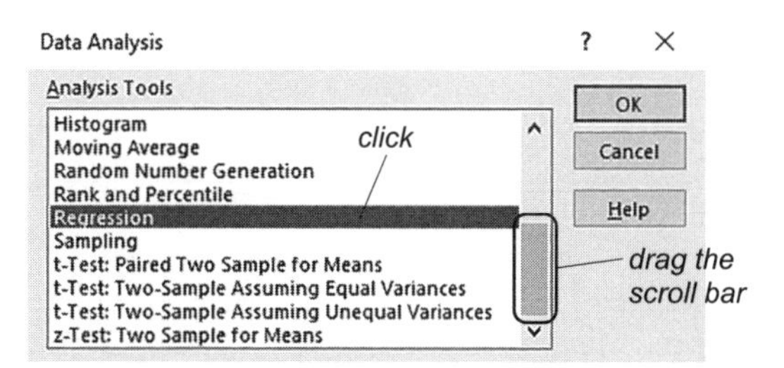

Fig. 13.1 Select the [Data] tab → [Data Analysis] → [Regression].

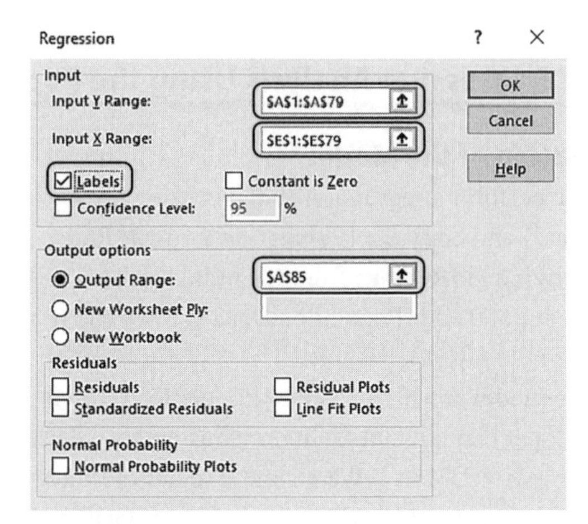

Fig. 13.2 The [Regression] box opens, and we input the data ranges. Input **A1:A79** in [Input Y Range] and **E1:E79** in [Input X Range]. Click [Labels] so that the first row is the field name (otherwise an error occurs). Click [Output Range] in [Output Option] and input **A85**.

and input **A85** (Fig.13.2). Click [OK], and the results are output from A85.

The results consist of three different parts, [Regression Statistics], [ANOVA], and the estimates of the coefficients and standard errors (Fig.13.3). In [Regression Statistics], R^2 (*R Square*), its square root (*Multiple R*), the value of s (*Standard Error*), adjusted R^2 (*Adjusted R Square*), and the number of observations (*Observations*) are presented. In ANOVA, *df* represents the degrees of freedom, *SS* is $\sum(\hat{Y}_i - \bar{Y})^2$, $\sum e_i^2$, and $\sum(Y_i - \bar{Y})^2$. $MS = SS/df$ is given for the regression values (*Regression*), residuals (*Residual*), and deviations of Y (*Total*). F is the ratio of MS in the residuals and regression, and *Significance F* is the p-value corresponding to F. (Later, we will learn about Adjusted R Square and F.)

Finally, the results of the estimation of the regression equation are presented.

	A	B	C	D	E	F	G	H	I
85	SUMMARY OUTPUT								
86									
87	*Regression Statistics*								
88	Multiple R	0.674066999							
89	R Square	0.454366319							
90	Adjusted R Square	0.447186928							
91	Standard Error	0.006579776							
92	Observations	78							
93									
94	ANOVA								
95		*df*	*SS*	*MS*	*F*	*Significance F*			
96	Regression	1	0.002739939	0.002739939	63.28759	1.34027E-11			
97	Residual	76	0.003290303	4.32935E-05					
98	Total	77	0.006030241						
99									
100		*Coefficients*	*Standard Error*	*t Stat*	*P-value*	*Lower 95%*	*Upper 95%*	*Lower 95.0%*	*Upper 95.0%*
101	Intercept	0.040233343	0.004353856	9.240854067	4.62E-14	0.031561889	0.048904798	0.031561889	0.048904798
102	Ln of GDP	-0.00383876	0.000482538	-7.955349715	1.34E-11	-0.004799818	-0.002877702	-0.004799818	-0.002877702

Fig. 13.3 Results of the simple regression analysis.

Coefficients and *Standard Error* are the estimates of the coefficients and standard errors, respectively. The sample regression equation becomes the following (standard errors are in parentheses).

$$Y = 0.04023 - 0.00389X \quad (13.16)$$
$$(0.004354) \quad (0.000483)$$

This result suggests that the population growth rate declines about 0.004% if the GDP per capita increases by 1%. In the table, *t Stat* (= Coefficients/Standard error), *p-value* (= two-tailed *p*-value), *Lower* 95%, and *Upper* 95%, the confidence interval with a 95% confidence coefficient, are given. The confidence interval with the confidence coefficient $1-\alpha$ is given by $\hat{\beta}_j \pm t_{\alpha/2}(n-2)\cdot s.e.(\hat{\beta}_j)$.

13.4.2 Test of Significance

Using the results of observation, we perform the test of significance. It is expected that the population growth rate decreases as the income level increases. The null and alternative hypotheses become

$$H_0 : \beta_2 = 0, \quad H_1 : \beta_2 < 0$$

The test statistic t is equal to *t Stat* = −7.955 in this case. Let the significance level α be 1%. The percent point is $t_{\alpha/2}(n-2) = 2.376$ (the percent point of the t-distribution, $t_\alpha(n-2)$ is obtained by **=TINV(2*α, n−2)**). Since $t < -t_{\alpha/2}(n-2)$, the null hypothesis is rejected. The regression equation is significant, and the income level is a significant variable that may affect the population growth rate. Note that the result of the test is the same as that of the correlation coefficient test in Section 12.3.

13.5 Multiple Regression Analysis

In the previous sections, we have studied a model that contains only one independent variable, known as simple regression analysis. However, in many cases, two or more independent variables may affect the dependent variable. For example, when we consider the consumption of a certain commodity, we can consider income, gender, the amount of assets, age, etc., as independent variables. When there are two or more variables, we call this multiple regression analysis.

The multiple regression equation contains the independent variables $X_2, X_3, \ldots, X_k$, and it satisfies

$$Y_i = \beta_1 + \beta_2 X_{2i} + \beta_3 X_{3i} + \cdots + \beta_k X_{ki} + u_i, \quad i = 1, 2, \ldots, n \quad (13.17)$$

in the population. $\beta_2, \beta_3, \ldots, \beta_k$ are unknown parameters and represent the pure effects of the independent variables, excluding the effects of other variables. u_i is an error term, and the following assumptions are made.

Assumption 1

$X_{2i}, X_{3i}, \ldots, X_{ki}$ are nonstochastic and take fixed values.

Assumption 2

u_i is a random variable and $E(u_i) = 0$, $i = 1, 2, \ldots, n$.

Assumption 3

Different error terms are uncorrelated, i.e., if $i \neq j$, $Cov(u_i, u_j) = E(u_i u_j) = 0$.

Assumption 4

The variances of the error terms are constant, and σ^2, i.e., $V(u_i) = E(u_i^2) = \sigma^2$, $i = 1, 2, \ldots, n$. We call this the homoscedasticity.

Assumption 5

An independent variable is not a linear function of other independent variables, i.e.,

$$\alpha_1 + \alpha_2 X_{2i} + \alpha_3 X_{3i} + \cdots + \alpha_k X_{ki} = 0, \quad i = 1, 2, \ldots, n$$

is true if and only if $\alpha_1 = \alpha_2 = \cdots = \alpha_k = 0$. We say that there is no perfect multicollinearity.

13.5.1 Estimation of the Model

Since the multiple regression equation contains k unknown parameters, $\beta_1, \beta_2, \ldots, \beta_k$, we estimate them from the sample. In the simple regression analysis, we use the least squares method. Namely,

$$u_i = Y_i - (\beta_1 + \beta_2 X_{2i} + \beta_3 X_{3i} + \cdots + \beta_k X_{ki}),$$

and we consider the sum of the squares,

$$S = \sum u_i^2 = \sum \{Y_i - (\beta_1 + \beta_2 X_{2i} + \cdots + \beta_k X_{ki})\}^2$$

and obtain $\beta_1, \beta_2, \ldots, \beta_k$ by minimizing S. For that purpose, we partially differentiate S with respect to β_j and put 0, we get k simultaneous equations

$$\partial S/\partial \beta_1 = 0, \ \partial S/\partial \beta_2 = 0, \ldots, \partial S/\partial \beta_k = 0.$$

Since the simultaneous equations are linear functions of $\beta_1, \beta_2, \ldots, \beta_k$, we can solve them. The least squares estimators $\hat{\beta}_1, \hat{\beta}_2, \ldots, \hat{\beta}_k$ give the solution of these simultaneous equations, also called the sample (partial) regression coefficients. Assumption 5 guarantees the existence of a unique solution. As the simple regression model, $\hat{\beta}_1, \hat{\beta}_2, \ldots, \hat{\beta}_k$ are the best linear unbiased estimators as determined by Gauss-Markov's theorem.

The equation

$$y = \hat{\beta}_1 + \hat{\beta}_2 x_2 + \hat{\beta}_3 x_3 + \cdots + \hat{\beta}_k x_k$$

and the estimators of $E(Y_i)$

$$\hat{Y}_i = \hat{\beta}_1 + \hat{\beta}_2 X_{2i} + \hat{\beta}_3 X_{3i} + \cdots + \hat{\beta}_k X_{ki}$$

are called the linear regression equation and the fitted value (regression value).

The variance of u_i, σ^2 is estimated from the residual $e_i = Y_i - \hat{Y}_i$ and

$$s^2 = \sum e_i^2/(n-k). \tag{13.18}$$

The sum of the squared residuals is divided by $n - k$ because e_i is estimated using $\hat{\beta}_1, \hat{\beta}_2, \ldots, \hat{\beta}_k$. We get $\sum e_i = 0$, $\sum e_i X_{2i} = 0$, $\sum e_i X_{3i} = 0, \ldots, \sum e_i X_{ki} = 0$ and the degrees of freedom are reduced by k. s^2 is an unbiased estimator of σ^2.

13.5.2 Goodness of Fit

The variation of Y_i, $\sum (Y_i - \bar{Y})^2$ is the sum of the portion of Y_i that can be explained by $X_2, X_3, \ldots, X_k$ and the portion that cannot be explained.

$$\sum (Y_i - \bar{Y})^2 = \sum (\hat{Y}_i - \bar{Y})^2 + \sum e_i^2.$$

The goodness of fit is given by R^2:

$$R^2 = 1 - \frac{\sum e_i^2}{\sum (Y_i - \bar{Y})^2} = \frac{\sum (\hat{Y}_i - \bar{Y})^2}{\sum (Y_i - \bar{Y})^2} \tag{13.19}$$

The positive square root of R^2, R, is called the multiple regression coefficient.

R^2 increases as the number of independent variables increases. If $k = n$, $R^2 = 1$. (If $k > n$, we cannot estimate the model.) However, if we use too many (unnecessary) variables, the model becomes less accurate (the variance of the estimator becomes large). If the numbers of variables are different, we cannot simply use R^2 to compare the goodness of fit. The adjusted R^2, $\bar{R}^2$, is used access the differences in the numbers of independent variables and is defined as

$$\bar{R}^2 = 1 - \frac{\sum e_i^2/(n-k)}{\sum (Y_i - \bar{Y})^2/(n-1)}. \tag{13.20}$$

$\bar{R}^2$ may not increase when k increases. Maximizing $\bar{R}^2$ is the same as minimizing s^2.

In multiple regression analysis, selecting the proper combination of independent variables (model selection) is very important. The previous studies imply that the penalty of $\bar{R}^2$ for increasing the independent variables is not large enough. One criterion that is widely used is the Akaike information criterion (AIC). In multiple regression analysis, it is equivalent to

$$\text{AIC} = \log_e \left(\frac{\sum e_i^2}{n} \right) + 2k/n \tag{13.21}$$

We choose a model that minimizes the AIC. In any case, it is important not to include unnecessary explanatory variables in multiple regression analysis.

13.5.3 Hypothesis Tests

a. *t*-Test

Using the estimator of σ^2, s^2, we can obtain the standard error of $\hat{\beta}_j$, $s.e.(\hat{\beta}_j)$. Here,

$$t_j = \frac{\hat{\beta}_j - \beta_j}{s.e.(\hat{\beta}_j)} \tag{13.22}$$

follows a t-distribution with degrees of freedom given by $n - k$ and $t\ (n - k)$. Therefore we can perform the hypothesis test of a single regression coefficient H_0: $\beta_j = a$, using the same method as in the simple regression analysis.

b. *F*-Test

In multiple regression analysis, we sometimes want to test a hypothesis concerning two or more regression coefficients. For example, suppose we give two medicines, A and B, to mice and evaluate their effects. Let the weight of the mouse be Y, and the amounts of the medicines given to the mouse be X_2 and X_3. Then, the null hypothesis of "neither of the two medicines affects the result" is

$$H_0 : \beta_2 = 0 \quad \text{and} \quad \beta_3 = 0.$$

The alternative hypothesis that "at least one variable affects the results" is

$$H_0 : \beta_2 \neq 0 \quad \text{or} \quad \beta_3 \neq 0.$$

In this example, when the hypothesis consists of two or more equations, t-tests of each coefficient are not sufficient, and it is necessary for us to perform the F-test as follows.

i) Assuming that H_0 is correct, we estimate the multiple regression equation (in the example above, the equation which does not include X_2 and X_3) and calculate the sum of the squared residuals, S_0.

ii) Adding all independent variables (assuming that H_0 is not true and adopting H_1), estimate the regression equation and calculate the sum of the squared residuals, S_1.

iii) Let p be the number of equations contained in H_0. Then

$$F = \frac{(S_0 - S_1)/p}{S_1/(n-k)} \tag{13.23}$$

follows the F-distribution with degrees of freedom given by $(p, n - k)$, $F(p, n - k)$ under the null hypothesis. The critical value of the test is the percent point of $F(p, n - k)$, with a significance level of α, $F_\alpha(p, n - k)$. We compare the test statistic F and $F_\alpha(p, n - k)$, reject the null hypothesis $F > F_\alpha(p, n - k)$, and accept it otherwise.

In particular, when we consider the null hypothesis that none of the independent variables explains Y,

$$H_0 : \beta_2 = \beta_3 = \cdots = \beta_k = 0,$$

and the alternative hypothesis

$$H_1 : \text{at least one of } \beta_2, \beta_3, \ldots, \beta_k \text{ is not } 0,$$

$p = k - 1$, $S_0 = \sum(Y_i - \bar{Y})^2$, $S_1 = \sum e_i^2$, $S_0 - S_1 = \sum(\hat{Y}_i - \bar{Y})^2$, and we calculate the value of F. This is called the F-statistic or F-value of the equation.

When the null hypothesis contains only one equation, $F = t^2$, the F-test becomes identical to the two-tailed t-test. In the F-test, we cannot perform the one-tailed test, where the alternative hypothesis is given by the inequality. Therefore, we use the t-test for the test of one coefficient.

13.6 Dummy Variables

In the regression analysis, we can use qualitative data as an independent variable using a dummy variable.

The dummy variable takes a value of 0 or 1. For example, when we express the effect of gender,

$$D_i = \begin{cases} 0 : \text{female} \\ 1 : \text{male} \end{cases}$$

Let Y be wages and X be the years of the experiment. Suppose that we analyze the wages of male and female workers. These wages are given by:

$$\text{female} : Y_i = \beta_1 + \beta_2 X_i + u_i$$

$$\text{male} : Y_i = \beta_1^* + \beta_2 X_i + u_i.$$

In this case, we can represent the wages of males and females using a single equation

$$Y_i = \beta_1 + \beta_2 X_i + \beta_3 D_i + u_i.$$

We can use the dummy variable in exactly the same way we use usual (quantitative) variables. For example, suppose that we want to know whether the wages of males and females are different or not. We test H_0: $\beta_3 = 0$.

The dummy variable is usually used in such analyses. But we can use it in different cases as well. Suppose that the initial wages at $X = 0$ are the same; however, pay raises that occur after that differ by gender:

$$\text{female} : Y_i = \beta_1 + \beta_2 X_i + u_i$$

$$\text{male} : Y_i = \beta_1 + \beta_2^* X_i + u_i.$$

Let $Z_i = D_i \cdot X_i$. We consider the model given by

$$Y_i = \beta_1 + \beta_2 X_i + \beta_3 Z_i + u_i.$$

The differences in pay raises between males and females can be tested using H_0: $\beta_3 = 0$.

In the case where both the initial wages and pay raises are different, we consider

$$Y_i = \beta_1 + \beta_2 X_i + \beta_3 D_i + \beta_4 Z_i + u_i.$$

The differences between males and females can be tested by the F-test of H_0: $\beta_3 = \beta_4 = 0$. (However, in this case, the result is identical to the result in which the regression equation is estimated for males and females separately.)

In the case of gender, the variable takes two different values. When the qualitative variable takes s different values, $A_1, A_2, \ldots, A_s$, we use $s - 1$ dummy variables, $D_{1i}, D_{2i}, \ldots, D_{s-1,i}$,

$$D_{1i} = \begin{cases} 1 & \text{if } A_1 \\ 0 & \text{otherwise} \end{cases} \quad D_{2i} = \begin{cases} 1 & \text{if } A_2 \\ 0 & \text{otherwise} \end{cases} \quad \cdots \quad D_{s-1,i} = \begin{cases} 1 & \text{if } A_{s-1} \\ 0 & \text{otherwise} \end{cases}$$

and analyze the data. Note that the number of dummy variables is $s - 1$, and we do not use the s-th dummy variable. If we use s dummy variables, we cannot estimate the model due to the perfect multicollinearity.

The one-way ANOVA model corresponds to the model in which all explanatory variables are dummy variables and we test the null hypothesis that none of the variables explain Y.

13.7 Multiple Regression Analysis Using the Population Data

13.7.1 Estimation of the Multiple Regression Equation

In Section 13.4, we performed a simple regression analysis in which Y is the [Population Growth Rate] and X is the natural logarithm of [GDP per capita]. Here, we add the population density and dummy variables representing regions (regional dummies) and perform a multiple regression analysis. As with GDP per capita, the minimum and maximum population densities differ by several digits, so we take the natural logarithm. Moreover, the population growth rate in Africa is considered to be high, so we use an Africa dummy that takes 1 if Africa and 0 otherwise.

The multiple regression equation becomes

$$(13.24) \quad \begin{cases} Y_i = \beta_1 + \beta_2 X_{2i} + \beta_3 X_{3i} + \beta_4 X_{4i} + u_i, \ i = 1, 2, \ldots, n \\ Y_i = \text{Population Growth Rate} \\ X_{2i} = \log_e(\text{GDP per capita}) \\ X_{3i} = \log_e(\text{Population Density}) \\ X_{4i} = \text{Africa Dummy} \end{cases}$$

Input **Ln of Density** in F1 and **=LN(C2)** in F2. Copy F2 to the data range. Next, we set the Africa dummy. Input **Africa Dummy** in G1 and **=IF(D2="Africa", 1, 0)** in G2, and copy it to the data range.

In simple regression analysis, select [Data] tab → [Data Analysis] →[Regression]. Input **A1:A79** in [Input Y Range] and **E1:G79** in [Input X Range]. Since the data ranges include the field names, click [Labels] so that the box is checked. Click [Output Range] in [Output options] and input **A105** (Fig. 13.4). Click [OK], and the results of the multiple regression analysis are presented from A105 (Fig. 13.5).

The results of the estimation are

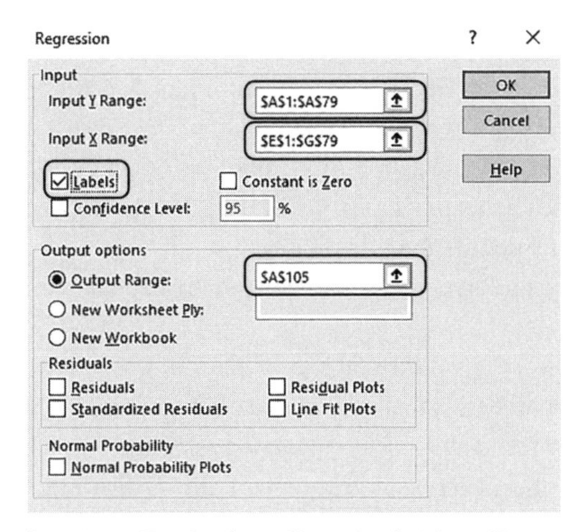

Fig. 13.4 [Data] tab → [Data Analysis] → [Regression]. Input **A1:A79** in [Input Y Range] and **E1:G79** in [Input X Range]. Since the data ranges include the field names, click [Labels] so that the box is checked. Click [Output Range] in [Output Option] and input **A105**.

	A	B	C	D	E	F	G	H	I
105	SUMMARY OUTPUT								
106									
107	*Regression Statistics*								
108	Multiple R	0.804043659							
109	R Square	0.646486205							
110	Adjusted R Square	0.632154565							
111	Standard Error	0.005367285							
112	Observations	78							
113									
114	ANOVA								
115		*df*	*SS*	*MS*	*F*	*Significance F*			
116	Regression	3	0.003898468	0.001299489	45.10902	1.0973E-16			
117	Residual	74	0.002131774	2.88078E-05					
118	Total	77	0.006030241						
119									
120		*Coefficients*	*Standard Error*	*t Stat*	*P-value*	*Lower 95%*	*Upper 95%*	*Lower 95.0%*	*Upper 95.0%*
121	Intercept	0.021250509	0.0059045	3.599035812	0.000574	0.009485535	0.033015482	0.009485535	0.033015482
122	Ln of GDP	-0.001603299	0.000551031	-2.909636354	0.004776	-0.002701251	-0.000505346	-0.002701251	-0.000505346
123	Ln of Density	-0.000887734	0.00048671	-1.823951121	0.072197	-0.001857524	8.20555E-05	-0.001857524	8.20555E-05
124	Africa Dummy	0.011055937	0.001934165	5.716128072	2.16E-07	0.007202028	0.014909845	0.007202028	0.014909845

Fig. 13.5 Result of multiple regression analysis.

(13.25)

$$Y = 0.02125 \quad - 0.001603X_2 - 0.000888X_3 + 0.01106X_4$$
$$(0.005905) \quad (0.000551) \quad (0.000487) \quad (0.001934)$$
$$s = 0.005367, \; R^2 = 0.6465$$

The standard errors are in parentheses.

13.7.2 Tests of Significance for Individual Coefficients

For $\beta_2, \beta_3, \beta_4$, we perform tests of significance for individual coefficients, meaning tests of whether the individual coefficient is 0 or not. Let the significance level α be 5%. Since the signs of $\beta_2, \beta_3, \beta_4$ are expected to be negative, negative, and positive, respectively, we perform a one-tailed test. The t-values are

$$t_2 = -2.910, \; t_3 = -1.824, \; t_4 = 5.716$$

We compare these values with the percent point of the t-distribution with degrees of freedom of $n - k = 74$ and $t_\alpha (n - k) = 1.666$. We reject the null hypotheses that the individual coefficients of $\beta_2, \beta_3, \beta_4$ are 0, and it is concluded that these variables may affect the population growth rate.

Next, we test the null hypothesis that none of the explanatory variables explain Y, given by

$$H_0 : \beta_2 = \beta_3 = \beta_4 = 0.$$

The alternative hypothesis is that at least one of them is not 0. Let the significance level α be 1%. The test statistic F is calculated as F, and $F = 45.109$. Since $p = 3$ and $n - k = 74$, the degrees of freedom of the F-distribution if (3, 74) and $F_\alpha (k - 1, n - k) = 4.058$. The null hypothesis is rejected, and it is concluded that these variables explain Y significantly. The percent point of the F-distribution is calculated by **FINV(*α*, k−1, n−k)**. *Significance F* is the p-value, and we can test using it. The null hypothesis is rejected if *Significance F* is smaller than α and it is accepted otherwise.

Finally, we test whether the two newly added variables, [Ln of Density] and [Africa Dummy], are meaningful variables or not by the F-test. The null hypothesis is

$$H_0 : \beta_3 = \beta_4 = 0.$$

The alternative hypothesis is that at least one of them is not 0. Let the significance level α be 5%. Under the null hypothesis, the explanatory variable is just [Ln of GDP], and the model used becomes the simple regression model we estimated in Section 13.4. The squared sum of the residuals is given in *SS* of the residual. From the results of the simple regression analysis, $S_0 = 0.003290$. When the null hypothesis is not correct, and the model contains all explanatory variables, the model used is the currently considered model, and $S_1 = 0.002132$.

Since $n - k = 73$ and $p = 2$, we get the F-statistic

$$F = \frac{(S_0 - S_1)/p}{S_1/(n-k)} = \frac{(0.003290 - 0.002132)/2}{0.002132/72} = 20.108$$

The number of degrees of freedom of the F-distribution is (2, 74), $F > F_\alpha(p, n - k) = 3.120$, and we reject the null hypothesis (Fig. 13.6).

	A	B
129	F-test beta2=beta3=0	
130	S0	0.003290303
131	S1	0.002131774
132	p	2
133	n-k	74
134	Test Statistic F	20.10794464
135	Significance Level	5%
136	Percent Point	3.120348511
137	Refect Null Hypothesis	
138		

Fig. 13.6 Result of *F*-test.

13.8 Exercises Using the Population Data

1. Do a simple regression analysis with Y as [Population Growth Rate] and X as [Ln of Density]. Test H_0: $\beta_2 = 0$ using the t-test. Choose a proper significance level and alternative hypothesis from now on.
2. As the regional dummies, calculate:
   i) Asia Dummy: 1 if Asia, 0 otherwise
   ii) Europe Dummy: 1 if Europe, 0 otherwise
3. Let Y be [Population Growth Rate] and let the explanatory variables X_2, X_3, X_4 be three regional dummies (Africa, Asia, Europe) and do a multiple regression. Test whether all dummy variables do not explain Y by the F-test. Compare the result with the result of Exercise 2 in Section 12.4 and check whether they both give the same result. (Hereafter, choose proper alternative hypotheses and significance levels.)
4. Let Y be [Population Growth Rate] and let [Ln of GDP], [Ln of Density], and three regional dummies be the explanatory variables X and perform a multiple regression analysis. Then do
   i) t-tests of the significance of individual coefficients
   ii) F-test of whether none of the variables explain Y
   iii) F-test of whether the three dummies do not explain Y
5. As already studied, AIC, which is widely used in model selection (the

selection of explanatory variables in regression analysis), is given by AIC $= \log_e (\sum e_i^2/n) + 2k/n$ in regression analysis. Find the model that minimizes AIC using the following steps.

i) Do a multiple regression analysis using all possible explanatory variables and calculate AIC.

ii) Eliminate the one variable whose absolute value for the t-value is the smallest, and calculate AIC.

iii) Repeat (ii) until only one explanatory variable remains.

iv) Choose the model that minimizes AIC.

References

○Statistical Theories and Methodologies

Anderson, D. R., D. J. Sweeney, T. A. Williams, et al. *Statistics for Business & Economics,* 14th ed. (*Essentials of Statistics for Business and Economics,* 9th ed.), Cengage, 2019.

Freedman, D. *Statistical Models: Theory and Practice,* Cambridge University Press, 2009.

Freedman, D., R. Pisani, and R. Purves. *Statistics,* 4th ed., W.W. Norton, 2007.

Spiegelhalter, D. *The Art of Statistics: Learning from Data*, Penguin Random House (Pelican), 2019.

Ubøe, J. *Introductory Statistics for Business and Economics: Theory, Exercises and Solutions*, Springer, 2017.

Weiss, N. A. *Elementary Statistics, MyLab Revision with Tech Update,* 10th ed., Pearson, 2016.

○Excel, General

Alexander, M., D. Kusleika, and J. Walkenbach, *Excel 2019 Bible*, Wiley, 2019.

○Excel, Functions and Formulas

George, N. *Excel 2019 Functions: 70 Top Functions Made Easy*, Amazon.co.jp, 2019.

McFedries, P. *Microsoft Excel 2019 Formulas and Functions*, Pearson, 2019.

○Excel, Macro and VBA

Alexander, M. *Excel Macros for Dummies,* 2nd ed., Wiley, 2017.

Alexander, M. and D. Kusleika, *Excel 2019 Power Programming with VBA*, Wiley, 2019.

Alexander, M. and J. Walkenbach. *Excel VBA Programming for Dummies,* 5th ed., Wiley, 2019.

Jalen, B. and T. Syrstad, *Microsoft Excel 2019 VBA and Macros*, Pearson, 2019.

○Excel, Data Analysis

McFedries, P. *Excel Data Analysis for Dummies,* 4th ed., Wiley, 2019.

Schmuller, J. *Statistical Analysis with Excel for Dummies,* 4th ed., Wiley, 2016.

Winston, W. L. *Microsoft Excel 2019 Data Analysis and Business Modeling*, 6th ed., Pearson, 2019.

Index

F

G

著者略歴

縄田和満　Kazumitsu NAWATA
東京大学大学院工学研究科教授
Professor in Graduate School of Engineering, University of Tokyo
一橋大学社会科学高等研究院特任教授
Specially Appointed Professor in Research Center for Health Policy and Economics, Hitotsubashi Institute for Advanced Study, Hitotsubashi University
Ph. D (Economics)

著　書 『Excelによる統計入門』(朝倉書店，1996年)
『Excelによる統計入門　第2版』(朝倉書店，2000年)
『Excelによる統計入門　Excel 2007対応版』(朝倉書店，2007年)
『Excelによる統計入門　第4版』(朝倉書店，2020年)　など多数

Introduction to Statistics Using Excel

定価はカバーに表示

2021年 8月 1日　初版第1刷

著　者　縄　田　和　満
発行者　朝　倉　誠　造
発行所　株式会社　朝　倉　書　店
東京都新宿区新小川町 6-29
郵便番号　162-8707
電　話　03(3260)0141
FAX　03(3260)0180
http://www.asakura.co.jp

〈検印省略〉

© 2021〈無断複写・転載を禁ず〉

中央印刷・渡辺製本

ISBN 978-4-254-12262-6　C 3041　Printed in Japan

JCOPY　<出版者著作権管理機構 委託出版物>
本書の無断複写は著作権法上での例外を除き禁じられています．複写される場合は，そのつど事前に，出版者著作権管理機構（電話 03-5244-5088，FAX 03-5244-5089，e-mail: info@jcopy.or.jp）の許諾を得てください．

東大 縄田和満著

Excelによる統計入門（第4版）

12243-5 C3041　A 5 判 208頁 本体2900円

初版刊行から20年以上版を重ねる定番テキストの最新改訂。文字の入力方法に始まり，表計算，グラフ作成，データ整理などExcel操作の基礎を学んだ後，記述統計量，二次元データの分析，推定・検定，回帰分析など統計学の基礎へ展開。

慶大 中妻照雄著

実践Pythonライブラリー

Pythonによる ベイズ統計学入門

12898-7 C3341　A 5 判 224頁 本体3400円

ベイズ統計学を基礎から解説，Pythonで実装。マルコフ連鎖モンテカルロ法にはPyMC3を活用。〔内容〕「データの時代」におけるベイズ統計学／ベイズ統計学の基本原理／様々な確率分布／PyMC／時系列データ／マルコフ連鎖モンテカルロ法

慶大 中妻照雄著

実践Pythonライブラリー

Pythonによる計量経済学入門

12899-4 C3341　A 5 判 224頁 本体3400円

確率論の基礎からはじめ，回帰分析，因果推論まで解説。理解してPythonで実践〔内容〕エビデンスに基づく政策決定に向けて／不確実性の表現としての確率／データ生成過程としての確率変数／回帰分析入門／回帰モデルの拡張と一般化

東北大 浜田　宏・関学大 石田　淳・関学大 清水裕士著

統計ライブラリー

社会科学のための ベイズ統計モデリング

12842-0 C3341　A 5 判 240頁 本体3500円

統計モデリングの考え方と使い方を初学者に向けて解説した入門書。〔内容〕確率分布／最尤法／ベイズ推測／MCMC 推定／エントロピーとKL情報量／遅延価値割引モデル／所得分布の生成モデル／単純比較モデル／教育達成の不平等／他

兵庫県大 笹嶋宗彦編

Pythonによるビジネスデータサイエンス　1

データサイエンス入門

12911-3 C3341　A 5 判 136頁 本体2500円

データの見方の基礎を身につける。サポートサイトにサンプルコードあり。〔内容〕データを見る／関係性を調べる／高度な分析（日本人の米離れ，気温からの売上予測，他）／企業の応用ケース／付録：Anacondaによる環境構築／他

関学大 羽室行信編

Pythonによるビジネスデータサイエンス　2

データの前処理

12912-0 C3341　A 5 判 192頁 本体2900円

データ分析のための前処理の基礎とビジネス応用の実例を学ぶ。サポートサイトにサンプルコードあり。〔内容〕前処理の意義／データの収集／実践（公的統計，マーケティング，ファイナンス，自然言語処理）／付録：Pythonの基礎／他

シード・プランニング 牛澤賢二著

やってみよう テキストマイニング［増訂版］

—自由回答アンケートの分析に挑戦！—

12261-9 C3041　A 5 判 192頁 本体2700円

知識・技術・資金がなくてもテキストマイニングができる！手順に沿って実際のアンケート結果を分析しながら，データの事前編集，単語抽出，探索的分析，仮説検証的分析まで楽しく学ぶ．最新のKH Coder 3に対応した待望の改訂版。

前首都大 朝野熙彦編著

ビジネスマンがはじめて学ぶ ベイズ統計学

—ExcelからRへステップアップ—

12221-3 C3041　A 5 判 228頁 本体3200円

ビジネス的な題材，初学者視点の解説，ExcelからR（Rstan）への自然な展開を特長とする待望の実践的入門書。〔内容〕確率分布早わかり／ベイズの定理／ナイーブベイズ／事前分布／ノームの更新／MCMC／階層ベイズ／空間統計モデル／他

前首都大 朝野熙彦編著

ビジネスマンが一歩先をめざす ベイズ統計学

—ExcelからRStanへステップアップ—

12232-9 C3041　A 5 判 176頁 本体2800円

文系出身ビジネスマンに贈る好評書第二弾。丁寧な解説とビジネス素材の分析例で着実にステップアップ。〔内容〕基礎／MCMCをExcelで／階層ベイズ／ベイズ流仮説検証／予測分布と不確実性の計算／状態空間モデル／Rによる行列計算／他

前都立大 朝野熙彦編著

ビジネスマンがきちんと学ぶ ディープラーニング with Python

12260-2 C3041　A 5 判 184頁 本体2800円

機械が学習する原理を，数式表現の確認，手計算，Pythonによる実装，データへの適用・改善と順を追って解説。仕組みを理解して自分のビジネスデータへの応用を目指す実務家のための実践テキスト。基礎数学から広告効果測定事例まで。

上記価格（税別）は 2021 年 7 月現在